WORLD
EATERS

WORLD
EATERS

HOW VENTURE CAPITAL IS CANNIBALIZING THE ECONOMY

CATHERINE BRACY

DUTTON

DUTTON

An imprint of Penguin Random House LLC
1745 Broadway, New York, NY 10019
penguinrandomhouse.com

LIBRARY OF CONGRESS CATALOGING-IN-PUBLICATION DATA

Names: Bracy, Catherine, author.
Title: World eaters : how venture capital is cannibalizing the economy / Catherine Bracy.
Description: New York : Dutton, [2025] | Includes index.
Identifiers: LCCN 2024025585 (print) | LCCN 2024025586 (ebook) |
ISBN 9780593473481 (hardcover) | ISBN 9780593473504 (epub) |
Subjects: LCSH: Venture capital. | Economics. | Food industry and trade.
Classification: LCC HG4751 .B74 2025 (print) | LCC HG4751 (ebook) |
DDC 332/.04154–dc23/eng/20241208
LC record available at https://lccn.loc.gov/2024025585
LC ebook record available at https://lccn.loc.gov/2024025586

Graphs on pages 22 and 28 courtesy of Jenna Van Hout

Printed in the United States of America
1 3 5 7 9 10 8 6 4 2

BOOK DESIGN BY LAURA K. CORLESS

The authorized representative in the EU for product safety and compliance is
Penguin Random House Ireland, Morrison Chambers, 32 Nassau Street,
Dublin D02 YH68, Ireland, https://eu-contact.penguin.ie.

For the Builders

CONTENTS

WORLD EATERS

INTRODUCTION

I t's hard to imagine a time and place that felt more optimistic and dystopian at once than San Francisco in the first half of the 2010s. It was the apex of the second tech boom, and the Bay Area was swimming in obscene amounts of money. Between them, Google, Facebook, and Apple were worth almost a trillion and a half dollars. A generation of newer companies like Uber and Airbnb, born out of the ashes of the Great Recession, were leading a second wave of tech behemoths that made their home in San Francisco rather than in the suburban office parks of Silicon Valley. Both Uber and Airbnb, and later Lyft, were the darlings of the era's "sharing economy," a term that with hindsight now sounds Orwellian but at the time spoke to the sense of naive possibility that many of us were eager to indulge.

I arrived in San Francisco right at this moment, dispatched from President Barack Obama's reelection campaign headquarters in Chicago, where I was working as a product manager on the campaign's technology team, to open a novel "technology field office." The tech field office would serve as an extension of the ambitious and innovative tech operation the campaign was spinning up in Chicago. Desperate for seasoned engineers and developers but unable to attract enough of them to Chicago, the campaign's leadership sent me to

California to do what organizers do: meet our people where they are. If they wouldn't come to us, we would go to them.

Being in the Bay Area around that time was exhilarating. We were squarely in the middle of the "yes we can" era. The Great Recession was ending, an unprecedented bull market was accelerating, and the tech industry finally seemed like it was going to deliver the society that the internet's techno utopian pioneers had promised.

By the middle of the decade, though, the bloom was coming off the rose. Five full years from the deepest point of the economic crisis, it was clear that the wealth being created by the booming recovery was accruing mostly at the top as inequality hit unprecedented levels. In the Bay Area the sense of haves versus have-nots was particularly stark, and tech bore the brunt of the blame. The cost of living was skyrocketing. Rents in the Bay Area grew by 50 percent in the 2010s, higher than in any other major metro in the country. And there were high-profile instances of tech workers and entrepreneurs doing and saying boneheaded, insensitive things, including one famous incident when a tech founder referred to those living on San Francisco's streets as "hyenas." Behavior like this cemented the impression for everyone else that the tech community—the companies and workers alike—were only in it for themselves.

Then came the buses. The economic recovery of the early 2010s coincided with the boom of cities. College-educated millennials who worked in the information economy, the "creative class" as Richard Florida termed them, had turned against suburbs and were seeking out the cultural richness and density that city living afforded. This meant that many of the twentysomethings flocking to the Bay to work at tech companies were putting down roots in San Francisco. Public transportation being what it is in car-centric California, the employers of this new wave of city-living workers lacked a reliable way to get them to the suburban corporate campuses in Silicon Valley.

The quickest solution was to set up private shuttle services to pick up workers in the city and ferry them down the Peninsula. In the eyes of the companies, not only was this a perk for their workers, who didn't have to deal with commuter traffic or navigate a less than ideal public transportation system, but the Wi-Fi-enabled buses allowed them to be productive during the couple of hours a day they were traveling to and from the office. And of course, there was the benefit to the environment of taking potentially thousands of cars off the road every day.

But the experience for everyone else was different. The hulking double-decker luxury coaches with blacked-out windows were a conspicuous nuisance on city streets that weren't designed to accommodate them. Adding insult to injury, they used public bus stops to pick up workers. The optics of well-off, mostly young, white tech workers stepping around those waiting for the city buses—many of whom were lower-income people of color who were not welcome on the tech shuttles—created a stark symbol of everything bad about tech in the city: entitled millennials and their employers segregating themselves from everyone else. Instead of using their privilege and influence to make the public transportation system better for everyone, they bought a way around it, even as they relied on public infrastructure to enable their private system.

By 2015, the civic dynamic in San Francisco had become truly toxic. It was common to see "fuck tech" graffitied around the city. I heard from many tech workers who had been spit on while walking around their neighborhoods. In 2014, at one demonstration against tech's incursion into the region, a protester climbed onto the roof of a Yahoo! shuttle bus and vomited down its windshield. There was even an incident where someone shot live ammunition at buses as they headed down a freeway between San Francisco and Silicon Valley.

But while the negative sentiment against tech was ratcheting up in San Francisco, the general sense across the Bay in Oakland, where I decided to put down roots after the campaign ended, was that tech gentrification was a San Francisco and Silicon Valley problem. The hassle of crossing the Bay to reach tech company offices gave long-time Oakland residents a false sense of security, that even if some gentrification happened, it would never be as bad as what our neighbors to the west were experiencing. The Oakland-based tech workers who were willing to deal with the longer commute were doing so because, for many of them, they valued what was special about "The Town," as it is known—its culture and diversity—and were willing to contribute to it rather than treat the city as a playground.

But even in our security across the Bay, it was hard to ignore what was happening in San Francisco, and Oaklanders watched with wariness, the way you might watch a wildfire move across the hills and hope the winds don't shift.

Then came a 2015 announcement that shocked my neighborhood: Uber was coming to town.

I knew almost immediately upon arriving in the Bay Area that if I were to live here for the long term, Oakland would be my home. It was a scrappy, diverse city with a chip on its shoulder, always in the shadow of its flashier neighbor across the Bay. It reminded me of Detroit, near where I grew up, and I fell in love at first sight.

I soon came to understand that Oakland's inferiority complex ran deep. But that dynamic was changing as we all watched San

Francisco morph into what seemed like an extension of the bland monoculture of Silicon Valley.

When Uber announced they were expanding to downtown Oakland, buying an old department store building and bringing 2,500 jobs, everything changed at once. Almost immediately after the news broke, a campaign called No Uber Oakland sprang up. The mayor, who hadn't been informed about the move in advance, hastily threw together a press conference in which she tried to walk a fine line between city boosterism (there was no doubt Oakland could use the tax base Uber would bring) and channeling the trepidation that many of her constituents felt.

As a newly die-hard Oaklander myself, I understood where the fear was coming from. At the same time, as a civic technologist who had spent my whole career working to make the internet a force for shared prosperity and more robust democracy, I was riveted by the dynamics.

Before I joined the Obama campaign, I had spent the first eight years of my career working at Harvard's Berkman Center for Internet & Society (now the Berkman Klein Center). Those eight years, 2002 through 2010, were some of the most important in the history of the internet. Mark Zuckerberg and a group of his Harvard roommates, one of whom was a student who spent time at the Berkman Center, started Facebook in 2004. What is thought to be the first podcast was recorded at the Berkman Center's conference room table. We incubated a community that grew into an organization called Global Voices, which pioneered the practice of "citizen journalism" and fostered democracy around the world. We hosted idealistic operatives from Howard Dean's internet-driven campaign and watched in awe as a Harvard Law graduate leveraged digital tools to become the first Black president of the United States. I had a front-row seat for all of it, and by the time I joined President Obama's reelection campaign, I had come to believe that the internet was going to be a force for democracy and opportunity in the world—maybe even the most democratizing technology in human history.

So when Uber came to town and the community revolted, I had to ask myself some tough questions. Uber was the most valuable privately owned company in the world. They were offering to bring thousands of high-quality jobs to a place where the median income was about $25,000 per year, but the community saw it as a threat instead of an opportunity. Why? Did it have something to do with Oakland's famously leftist political sensibilities? Was there something specific to the economic context of the place or time? Or were the characteristics of the tech sector itself causing its growth to create harm for everyone else? Most important, what would it take for tech's growth to be a rising tide that lifted all boats?

I decided to devote myself to answering those questions. I cofounded an advocacy organization, TechEquity Collaborative, in an attempt to bend the trajectory of the tech industry's impact on the economy—as well as the corrosive effect that tech-fueled inequality was having on the civic dynamic. Maybe we could turn Uber's presence in Oakland from a crisis to an opportunity, I thought. Instead of trying to stop Uber from coming, we could engage the tech industry to be better corporate citizens, more like the auto companies of my midwestern youth who saw their success as intrinsically tied to the communities around them. We would engage both tech workers, in their role as citizens of this place, and companies, who could use their immense power to advocate for policy changes that would ensure that, as their industry continued to grow, it would create broad-based opportunity for the entire population. After all, it became clear to me very quickly that there were deep structural flaws in

California's economy that would have led any booming industry, tech or otherwise, to cause the kind of displacement and inequality we were seeing at the time.

But, as I learned over the course of the several years I tried—and mostly failed—to get tech companies to use their power in this way, there was also something unique to this particular industry that was making things worse than they might have been.

It wasn't obvious to me at first, and I became frustrated by the intransigence I was encountering. The *people* in these companies seemed as though they wanted to work with us. But when it was time for the *companies* to take action, the conversations almost always came to a halt. Why wouldn't they, for example, support a ballot measure that corrected an inequity in the tax code that was making their tax bills higher, and at the same time would fund improvements to the public education system that would train the workers they would need to ensure their future success? Or support affordable housing or workforce development initiatives that made it easier for them to hire much-needed talent? Eventually, after badgering enough Silicon Valley insiders at cocktail parties and networking events, it dawned on me that the irrational thoughtlessness I perceived was actually a perfectly rational reaction to the system in which these companies operated. That system incentivized these companies into thinking on short-time horizons and to view growth in terms of breadth, not depth.

That system was venture capital.

I'm a little embarrassed at how long it took me to reach this conclusion, given how ubiquitous and inextricable venture capital is in Silicon Valley. But I'm not the only one for whom this epiphany was a long time coming. If you follow the public debate about the various harms that tech companies have inflicted on society, and what regulators should do about it, you'd think these companies sprang fully formed from the minds of their founders. This lack of a robust understanding of the VC-created incentive structure that drives the tech industry is leading many policymakers to propose solutions that won't sustainably address the problem. The general consensus seems to be that the challenge with tech is that the companies got too big, and the way to address it is by wielding antitrust laws to make them give up their power. Senator Elizabeth Warren, one of Big Tech's most vocal critics in Washington, has said that the way to address the harms that tech has caused is to "enshrin[e] strong antitrust principles into new legislation" and "reviv[e] serious antitrust enforcement at both the [Federal Trade Commission] and [the Department of Justice]."

More robust application of antitrust laws is certainly a critical element of a larger strategy to hold the tech industry to account. But these companies didn't become as powerful as they are by accident. Prolific venture capitalist and aspiring political kingmaker Peter Thiel famously said "competition is for losers." He speaks for a whole system that has created the incentives, imperatives, and culture out of which these dominant companies arose. Venture capitalists benefit from getting to operate in the background while the companies they invest in capture the public's eye. And since VCs are investing in private companies at their earliest stages, the full role venture capitalists play in the tech industry isn't entirely transparent. But after reporting for this book, it's clear to me that venture capitalists have much more influence than we give them credit for. If we are to reckon with what venture capital hath wrought—not just through Alphabet, Amazon, and Meta, but through the next generation of behemoths that VCs are currently nurturing—we have to understand it, and account for the role it plays in shaping the tech industry and the economy at large.

Over the past ten years, a genre of books and other media has emerged to expose egregious and often salacious harms, abuses, and excesses that are too common in the tech industry. I don't mean for this book to be one of them. In fact, the more time I spent researching this book, the more I came to believe in venture capital's proper role in the economy. There is no doubt that VC has created huge benefits for society. Leaving aside the economic gains that venture capital has created, many venture-backed companies—companies that may not have been able to find financial support anywhere else— have unquestionably improved the quality of life for hundreds of millions of people around the world. Genentech, one of the biggest venture capital successes of all time, is responsible for the mass-market production of human insulin. Google made the vast array of knowledge on the internet available for free to anyone who could get online. Moderna developed a Covid vaccine with breathtaking speed, paving the way for the world to emerge from the pandemic and likely saving millions of lives in the process. All these breakthrough technologies were funded by venture capital.

But the story also has a dark side. At the same time venture-backed companies were improving quality of life, they were also undermining it. Social media companies like Facebook and Twitter provided the tools that allowed people to take down dictators— and then put those same tools into the hands of authoritarians who used them to spread disinformation and savagely oppress their people. Lyft and Uber created a critical extension of transportation infrastructure—but they also increased traffic, undercut public transit, and eroded worker protections for millions of people. Juul purported to be creating a safe alternative to cigarettes—only to get another generation hooked on nicotine.

I don't think it has to be this way. In the following pages, I hope to show that what is broken about the tech industry isn't primarily a function of the technology itself, or even the companies that build and sell it. The brokenness goes much deeper than that, into the economic system that made those companies and tools possible in the first place: venture capital.

Venture capital creates this brokenness in two ways. *First*, the venture capital industry has become unwaveringly committed to an investing approach that demands venture-backed startups pursue hyper maximalist growth at breakneck pace. This methodology—embodied by the creed made famous by Facebook: "Move fast and break things"—forces companies to make thoughtless and often irresponsible decisions that result in negative social and economic outcomes, for which the rest of us bear the cost.

Second, and arguably more crucially, the venture capital approach to investing has crowded out other forms of capital that could support sustainable startup development. What results is an unhealthy monoculture where only one kind of company can be successful. Entrepreneurs with great ideas that don't fit the mold of venture-scale growth, or who aren't willing to compromise their values at the altar of investor returns, are often either left to die or forced to morph to fit the accepted paradigm. In this way, venture capital has deeply distorted the innovation ecosystem and has, I argue, killed more value than it has created. This is especially true in areas of the economy where the venture approach is ill suited for the market. What works to spur innovation for software and high-technology companies can be a disaster when applied to infrastructure-heavy sectors like housing and clean energy.

Throughout this book, I highlight stories from entrepreneurs that point out exactly the ways in which venture capital perpetuates this harm: forcing scale and speed on companies that don't have many other viable options to fund their startup's growth. One of these companies is LocalData.

After the Obama campaign ended in 2012, I wanted to continue working on tech projects that could improve our civics. So I took a job at a San Francisco–based nonprofit named Code for America, whose mission is to improve the delivery of government services by applying Silicon Valley techniques, like user-centered design and agile software development that came out of tech startup culture. Code for America's flagship program was at that time a fellowship that sent earnest young technologists and designers into city governments across the country for a year to embed these practices in government operations.

Alicia Rouault, an urban planner who had just finished a master's program at MIT, was a member of the 2012 fellowship class. Alicia and two other fellows were deployed to Detroit at a time when the city was still experiencing the fallout from the Great Recession. The city, on the brink of declaring bankruptcy, was responsible for managing infrastructure built to support two million people spread across 139 square miles. But by 2012, the population had plummeted to around 700,000, dramatically reducing the city's tax base and leaving whole city blocks abandoned. With an estimated 40 percent of the city's streetlights out, Detroit was literally struggling to keep the lights on. Rouault and her team were charged with finding ways to use technology to improve data collection about the vast number of properties that were falling into blight, so that the city could better prioritize their upkeep and redevelopment.

Rouault and her team applied user-centered design methods to learn about the city's pain points from city workers and residents, and they realized that the analog data-collection method the planning department used was out-of-date and inaccurate, leading to major problems for the city. Street teams were collecting information on paper, and then handing over that paper to other workers who needed to decipher and manually enter it into a database. Unreliable tax parcel data meant Detroit had uncertainty about how many vacant properties the city had on its rolls, which in some cases led to the demolition of the wrong houses, and billions owed in back taxes (a major contributing factor to the city's bankruptcy).

Rouault and her teammates realized they could use what was at the time emergent mobile technology to build an app that allowed the street teams to enter data directly into the database from their phones and attach it to the city's tax parcel data, saving time and avoiding transcription errors. The tool was a hit, winning a major grant from a local foundation that allowed the team to expand their pilot beyond Detroit to other cities facing similar challenges. They called the tool LocalData.

As they began to prove out LocalData's use case and refine the technology, Rouault and her teammates realized that their project, which up to this point they viewed as a public good and not a profit-making venture, was in high demand from other municipalities. They figured that turning LocalData into a business was a way they could increase the tool's impact and sustainably expand it to more places. "We decided to start a company and not a nonprofit because we thought it had a substantial government market we wanted to serve," Rouault told me recently. The team soon entered two Bay Area startup accelerators that targeted civic-minded founders, one at Code for America, which launched to help build an ecosystem of companies whose main customer was government, and one called Matter.vc, which focused on media and journalism startups.

LocalData was a success from the start. The company was small, with fewer than ten staff members, but it was profitable—a rarity in the startup world—from almost the beginning. LocalData experienced 30 percent growth in sales in just its first year. Rouault and her cofounders were encouraged by the mentors they met through their accelerators to go out and raise venture capital so they could scale their business. "We had sustainable revenue, but we didn't have enough to gain more customers quickly," she told me. So they turned their attention from serving their existing customer base and went out to pitch to VCs.

What they heard from those VCs was discouraging. "A lot of the reactions we got were 'this is cool and it seems like you have traction, but the market isn't large enough,'" remembered Rouault. To venture capital investors, LocalData's decision to focus on government customers limited the size of the total addressable market (or TAM, as it is known). In order to convince investors that they had venture-scale potential, they would have to expand their scope. Many VCs suggested that they should shift their focus to customers in the real estate sector rather than governments. "It wasn't a critique of the product or even the business model, it was like 'you need a bigger and better market,'" said Rouault. "So we pivoted."

That pivot was a mistake. While some real estate companies were excited about incorporating data into their workflows, it turned out the majority of real estate firms were hard to convince. At the time, real estate was still a very analog sector, based on relationships and not driven by data. LocalData's potential customers were deeply skeptical that an app could replace their well-worn systems. Try as they might, Rouault and her teammates were told they were "too soon" for where the real estate sector was at that time. Compounding the problem, their ill-fated attempt to woo the real estate sector took them away from growing the government business that was already working. "It actually pushed us further and further away from the actual use case we had validated with real people," she said. In her mind, venture capitalists had pushed her to a solution in search of a problem. Rouault's experience with turning the company's attention away from the customers who she already knew were finding value from their product to "bending over backwards to meet the VC need," as she put it, soured her on the startup scene and its venture-infused culture.

When I first heard this story, back in 2014, Rouault was telling me how her team was going to have to shut down LocalData—essentially that the company was failing—because they couldn't raise any money from investors. "But . . . isn't the company profitable? Aren't you growing?" I remember asking her with genuine confusion. Yes, she told me, but if they couldn't convince venture capitalists that the market opportunity was big enough, the company had no future. She was told "[LocalData] is a lifestyle company," a pejorative term in VC that refers to businesses that are seen as too small to be worth the effort.

We'll revisit Rouault's experience in chapter 2, to examine in more detail the forces behind the demise of her company. But the questions she asks herself now, ten years later and having given up on her startup, point to the challenges created by venture capital's dominance of the capital markets for innovative startups. "Could our team of advisors have connected us to other forms of capital that were more appropriate for what we were trying to do?" she wondered. "Why didn't we look into more traditional sources of capital, like bank loans or small business grants? It's not like [the act of] financing companies was invented by the tech sector."

The essence of the problem Rouault and LocalData faced was that they couldn't connect to the right kind of capital for the type of business they were building. The market that LocalData was serving was too small to achieve the grand slam growth that venture requires, but it was a market with customers with a real need who had demonstrated that they were willing to pay for a solution. Even so, Rouault and her partners were not pushed to explore other options available to them. As far as she was concerned, and according to the people who were advising her, venture capital was the most viable path, and it was taken for granted that she should try and pursue it. Even more than that, amidst the gauzy heyday of the mid-2010s startup scene in the Bay Area, Rouault had internalized the notion that raising venture capital was *the* signal of success even if it wasn't the best fit for LocalData. And venture capitalists were all too willing to try and force her to fit that mold.

To understand how Rouault found herself in the position of feeling like a failure, you have to understand what venture capital is and where it came from. The first section of this book tells the story of venture capital's evolution, explains its structure, and outlines the specific elements of the venture capital methodology that make it so risky. It also introduces some of the key players—the institutional investors, known as limited partners, who pump money into venture capital funds; the venture capitalists themselves, also known as general partners, who distribute that money to startups; and the entrepreneurs who put it to use to grow their businesses. These players operate within an incentive structure that shapes their behavior. The first section of this book also takes a closer look at how the venture capital methodology plays out on the ground for real startups like LocalData, demonstrating in personal and sometimes heartbreaking terms how venture capital warps the innovation ecosystem.

The book's second section focuses on how venture capital affects the economic well-being of the rest of us. By taking a close look at the two areas of the economy that have the most impact on the economic prospects of everyday people—housing and labor—I show how venture capital is exacerbating inequality.

Despite the bad reputation the tech industry has developed for itself, you don't have to go far to find smart, dedicated entrepreneurs and investors who are doing inspiring work developing models for capital distribution and corporate formation that challenge the dominance of venture capital. The third section of the book tells their stories and points to ways that we can encourage a healthier innovation ecosystem—not only by putting guardrails around the excesses of venture capital but also by amplifying and supporting the work of these pioneers.

This book is meant to be about the failure of systems, not people. But systems don't fix themselves; they change only when people do. I hope this book is a call to action for those who have a role to play in creating an innovation ecosystem that produces not just financial gain for a few but broad-based economic growth that benefits everyone. Along the way, we might just have a better chance of solving our most intractable challenges to boot.

What is it exactly that these stakeholders can do?

- Founders must come to understand that raising venture capital isn't the mark of a successful entrepreneur, and that the most critical part of getting their business off the ground is finding investors whose incentive structures align with their goals for the company.

- Regulators should include venture capital in their analysis as they seek to create more accountability for the tech sector, and look at how they might institute policy to ensure both that venture capital is playing the correct role in the financial system and that all types of innovative companies have access to the capital they need to succeed sustainably.

- Venture capitalists should challenge Silicon Valley's orthodoxies and pursue other approaches to creating investment returns than the single one to which they have almost all been committed for the last several decades. I know most of them believe it can't be done, but isn't the mark of a good venture capitalist that they believe in doing things that other people think are impossible?

- Most important, limited partners—the wealthy individuals and institutions who provide the capital that venture investors distribute—should start using their influence to shape the venture capital industry. Many of the institutions whose money is flowing into venture capital claim to care about world-positive outcomes. But they often take a passive approach to overseeing their portfolios, delegating responsibility to fund managers whose compensation packages incentivize them to seek the highest possible returns at all costs. If just some of these limited partners decided to support a different approach to startup investing, it could create a wave that shifts the economy.

For the rest of us—parents concerned about how tech is influencing our children, advocates who fight for a fairer economy, and citizens who care about the health of our democracy—I hope this book provides a clearer picture of how the tech sector works so we can lobby our representatives for changes that will actually fix the problems we are trying to solve.

Above all, this book is an invitation to a conversation about how we build a future where venture capital is a force for good, enabling human flourishing rather than eroding our well-being.

In 2011, just as the second tech boom was gaining steam, Marc Andreessen, one of the most prolific and influential venture capitalists in history, published a now-famous essay in *The Wall Street Journal* titled "Why Software Is Eating the World." In it, he writes:

> *More and more major businesses and industries are being run on software and delivered as online services—from movies to agriculture to national defense. Many of the winners are Silicon Valley–style entrepreneurial technology companies that are invading and overturning established industry structures. Over the next 10 years, I expect many more industries to be disrupted by software, with new world-beating Silicon Valley companies doing the disruption in more cases than not.*

I hope by the time you finish reading this book, you understand why I think he is wrong. Or, more precisely, I hope you understand *how* I think he is wrong. It isn't software that disrupted those industries. It was the economic forces that were conjured and unleashed by investors, not engineers, that drove Silicon Valley's incursion into every corner of the economy. Yes, the nature of software, which is vastly cheaper to build than physical goods and can be delivered to almost anyone in the world almost instantly, enables the world-eating virality of Silicon Valley. But on its own, the software is just a tool. Its value, its ability to exploit or empower, derives from the economic system out of which it is born.

Those of us who care about ensuring that technology is enabling human flourishing, rather than undermining it, have for too long ignored or simply been unaware of the role of venture capital in our work. I didn't see it at first, when Uber was moving into Oakland, even as I wondered how our economy had become so polarized. I don't think a lot of regulators—in Washington, DC, Brussels, and elsewhere—see it either. Our narrow focus on the technology itself instead of the economic forces that surround it has allowed venture capital to largely avoid scrutiny during the rise and fall of the second tech boom, even though for every megalomaniacal founder who spectacularly crashed and burned, there were a dozen investors eagerly handing them the matches. Tech company CEOs are routinely trotted out in congressional hearings where they are berated for the harms their companies are seen to be creating. None of the venture capitalists who backed those companies get the same invite. If we don't pay the same scrutiny to those who made the rise of these companies possible—and built unfathomable wealth in the process—we have little hope of fixing what we believe tech has broken.

Venture capital fell into a lull at the end of 2021 just as interest rates began to rise. As we'll discuss in more detail in chapter 1, the end of the era of free money forced many venture capitalists to revisit the growth-at-all-costs approach. "The expectation for what a good company is, and how much risk you can take and how profitable you need to be and whether or not you are cash flow positive, those rules today are dramatically different than they were three, four years ago," said the famous venture investor Bill Gurley in 2024.

But, fueled by the swift emergence of generative AI and signals from the Federal Reserve that interest rates may be on their way back down, there are signs that the retrenchment to austerity may be ending. A survey by the Kauffman Fellows of venture capital sentiment in early 2024 showed that over half of venture capital fund managers believed economic conditions would improve that year, and over half of them expected to deploy more capital into startups than they had in 2023. If they're right, will Gurley's view, that venture capital investing has become inherently less risky, hold? Or is this just the beginning of another cycle of unfettered speculation?

While the original promise of the internet was that it would lower barriers to entry to all aspects of our economy and society, the industry that has grown up on top of it during the last two decades has taken us in the opposite direction. I'm still a believer in the original potential of the internet: that it can be the most democratizing communications platform in human history. Its import is clearer now than ever. But if we're going to realize that original promise, it isn't enough to simply hold accountable the companies themselves, or create guardrails around the tools they build. We have to understand and reckon with the economic structure that underlies them, that incentivizes them to become monopolies, that crowds out more sustainable options, and that is at this very moment investing in the next generation of unaccountable behemoths that will enforce their will on society—whether it's good for us or not.

1

THE METHODOLOGY

HOW VENTURE CAPITALISTS THINK

"Venture capital is not even a home run business.
It's a grand slam business."
–Bill Gurley, investor, Benchmark Capital

Over time, venture capital has come to be governed by a law of nature known as the power law. The power law is a naturally occurring phenomenon defined by the distribution of values within a certain type of dataset. Unlike in a normally distributed dataset, where most of the values are clustered around the average, in power law distributed sets, a tiny number of outliers can drag the average up all by themselves. Power law distributions have long tails, representing a large number of small values and a small number of very large values.

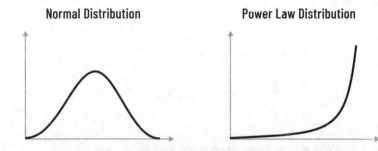

Normal Distribution   Power Law Distribution

Power laws can be observed in almost every corner of nature and society. The intensity of earthquakes, for example, is power law distributed. Most earthquakes, which occur with much more frequency than humans ever notice, don't even register on the Richter scale. But a small handful result in disastrous outcomes, leveling cities and defining how scientists relate to the phenomenon. Human populations are distributed according to a power law. Most people live in a tiny handful of very large cities, with the rest of the earth's population spread among many more smaller locales. Activity on the internet tends to be power law distributed too: a small number of sites on the web account for the majority of its traffic.

Venture capital funds, it turns out, also appear to be power law distributed. Venture capital funds are made up of a collection of companies, called a portfolio, usually a few dozen or so per fund. These companies are in early stages of development, and as such, there is much more uncertainty about their prospects. Unlike the companies funded by private equity, venture capital's more mature and boring older brother, the startups in a venture capital portfolio are, at least in theory, doing something interesting and innovative, often employing new technologies if not creating them. These characteristics lend themselves to a dynamic where there are lots of companies that fail to deliver on their vision but also a handful of world-changing breakthroughs that create enormous value. This end result, when these companies are compiled into a portfolio, is an overall fund value that is quite lucrative, even though only a small fraction of the companies within the fund are driving most of the value creation.

The relationship between the power law and venture capital has evolved over many decades, and in fact, as Tom Nicholas points out in his seminal history of venture capital, *VC: An American History*, has its roots in a much older industry. Whaling in nineteenth-century New England was, much as the tech industry is today, both highly lucrative and highly risky. Investors understood that if a whaling vessel had a successful venture, it could make them a huge return in exchange for a relatively small chunk of up-front capital. This was a big if. Whaling was highly unpredictable and very dangerous. It could take years for them to return, if they returned at all (many of them were lost at sea), and there was no guarantee that even if they did return, they would do so with a big enough catch to turn a profit. And yet, when they did succeed, they tended to succeed big.

Soon enough, savvy middlemen emerged who realized that if you spread around investors' money to multiple ventures, the chances of making a profit were much improved. If you aligned the incentives of the investors and the ship's crew by compensating the crew with a cut of the ship's haul instead of a regular salary, the chances of success went up even further. You could improve your odds even more by making bigger bets on the most successful ship captains.

The whaling industry died out by the end of the early 1900s, but it planted a seed for what would become a persistent idea in American capitalism.

It would be several decades before the techniques used by investors in whaling made their way to the world of high technology.

Prior to World War Two, there were very few sources of capital available to American entrepreneurs who wanted to bring risky innovations to market. Traditional banks were far too conservative to take the kind of risk necessary to support these startups, and fallout from the Great Depression limited the appetite for them to do so. Wall Street investment mainly went to support mature industries, making it very difficult for new technologies to emerge that could jump-start new sectors. More often, new innovations were fostered in the research and development departments of established firms, which made it harder for new entrants to compete. Breakthrough entrepreneurs in the early part of the twentieth century, like Thomas Edison and Henry Ford, had to rely on informal networks of wealthy people who treated early-stage investing more like a hobby than a business (a precursor to today's angel investors). Their ultimate success was almost an accident—a product of sheer force of will and being in the right place at the right time with access to the right networks.

As World War Two ended, and the Cold War began, there was increased urgency to commercialize the myriad technological innovations induced by the war, both to spur postwar growth and to make an ideological point about the power of capitalism. In 1946, a group of civic leaders and businessmen who were increasingly concerned with the lack of risk capital and associated innovation in the American economy decided they needed to take matters into their own hands. Led by a French émigré and Harvard Business School professor named Georges Doriot, they founded American Research and Development Corporation (ARD) to attract risk capital to innovative businesses. Their aim was to demonstrate that there was a way to earn reliable returns by investing in risky startups. Doriot and his partners hypothesized, much as the whaling agents of the previous century had, that to make investors comfortable with these high-risk investments, they should take a portfolio approach, creating exposure to multiple companies at a time. Nicholas quotes Merrill Griswold, one of the founders of ARD, as saying: "It is very risky to put money into a brand-new project. Some of them are bound to fail. But if you secure diversification, by buying fifteen or twenty of those indirectly through a special company, it does not matter that four or five of them may fail because the others, the hope is, will more than make up for it."

It took a while for ARD to prove this hypothesis correct, but when it did, it created a template for the generation of venture capitalists that would define the sector in the second half of the twentieth century. Eleven years after its launch, ARD invested in a small company called Digital Equipment Corporation (DEC) that made components for computers. DEC's founders, engineers who built some of the first computers at MIT and IBM, devised a way to lower the cost of computing, thereby providing access to computing power to a broader set of customers. Their products were so successful that by the end of the 1960s they were second only to IBM in the computing industry. ARD's $70,000 investment in DEC in 1957 (about $780,000 in 2024 dollars) was worth $230 million (over $2 billion in 2024 dollars) by 1968—a three thousandfold return, a demonstration of the power law in full effect. This one investment out of the dozens that ARD made in its entire twenty-five-year existence was responsible for doubling the firm's overall lifetime return. Without DEC, ARD wouldn't have outperformed the S&P composite index over the same period. With it, ARD beat the S&P by more than three percentage points.

THE RISE OF THE LIMITED PARTNERSHIP

Those numbers got the finance world to pay attention, and other investors who had fluency with the technology industries that were beginning to emerge in the vicinity of research hubs like Cambridge, Massachusetts, and Palo Alto, California, felt a wind at their back with the success of ARD's investment in DEC. New venture investment firms, such as Draper, Gaither & Anderson; Greylock Partners; and Venrock Associates, all emerged in ARD's wake.

But despite the influence ARD had on the venture capital industry, the quirks of its corporate structure created significant drawbacks. While ARD was investing in private companies, the fund through which it made the investments was publicly traded. This subjected it to a number of constraints that modern venture capital is not bound by. For example, the investing partners who worked for ARD could not receive equity stakes in their portfolio companies, which meant they didn't get a piece of the upside if one of their investments performed really well. This made it hard for ARD to compete for the best investor talent, because they couldn't offer a compensation package better than any other Wall Street salary. ARD also had to manage investor expectations for access to liquidity. Mid-twentieth-century investors often expected to be paid regular dividends on their shares. But the companies in the ARD portfolio wouldn't be making profits for years, if not decades. This illiquidity presented a problem that ARD solved by extending debt to its portfolio companies alongside equity investments, so that they could pass the monthly interest payments on the debt from the companies to their investors in the form of dividends. This debt burden on the companies weighed them down, making it harder for them to achieve the outsized scale that would deliver big returns in the long run.

Ultimately, all these factors taken together led to ARD's demise. The fund wound down in 1973 in the face of stiff competition from the upstart funds that ARD had inspired.

In many ways, those new firms were set up to replicate ARD's methodology: taking a portfolio approach to investing in companies, acting as hands-on advisers to the founders to build their management capacity, trying to identify risky bets that had huge upside potential and could drive power law returns. But they diverged from ARD's model in one critical way: they had stumbled upon a corporate structure—the limited partnership—that freed them from the oversight and overhead that ARD had to endure.

Limited partnerships are a type of organization designed to allow multiple actors to participate together in the ownership of a set of assets. The limited partnership acts essentially as a pass-through entity, allowing investors (the "limited partners," or LPs) to provide capital to general partners (for our purposes, the general partners are the venture capitalists) who organize it into a collection of funds that make investments in assets (for our purposes, these assets are startups). The limited partnership structure was not a new invention in the early 1960s when these firms adopted the model, but it did feel like one. First implemented in the United States in the early 1800s, it had more recently gained traction in the field of oil and gas speculation.

Perhaps the biggest advantage of limited partnerships is that, unlike the structure of ARD, they are private entities not subject to onerous oversight. Limited partnerships are required to register with the Securities and Exchange Commission (SEC), and general partner firms must register their funds when they create them, but

Limited Partnership Structure

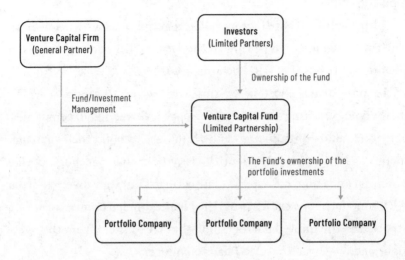

otherwise the parties in the limited partnership are accountable only to one another. The partnerships are "limited" in that the LPs are protected from any liability arising from the actions of the general partners or the companies in which they invest. The worst the limited partners can do is to lose their original investment. They cannot be held responsible if, say, one of the companies they invest in commits fraud or mishandles customer data. This limited liability made limited partnerships appealing to the kinds of investors who were skittish about the risk associated with untested startups—something that hampered ARD.

Limited partners are typically sophisticated investors, including wealthy families and institutional investors like university and private foundation endowments and, more recently, pension funds and insurance companies. As such, they are much more comfortable making longer-term investments than the public market investors who made up ARD's shareholders. This allows venture capital firms to invest entirely through equity rather than debt, freeing the portfolio companies from the strictures of making debt service payments. (Although there is such a thing as venture debt, and some startups do take out these loans alongside equity investments.) Equity investing also enabled limited partnerships to align the incentives of all the stakeholders involved, much as the arrangements used in the whaling sector aligned incentives across the investors, the agents, and the ship's crew. In the limited partnership structure employed by most venture capital firms today, general partners are paid through a combination of a flat fee (usually 2 percent of the total value of the fund per year) and a cut of the gains that their investments make (this varies, but 20 percent is standard). This structure is known as "two and twenty," and while the numbers differ slightly, depending on the fund, it has evolved into a remarkably consistent—and quite lucrative—compensation model (and one that, as we'll discuss later, is at the root of many of the harms of the modern venture capital system).

One of the biggest challenges facing ARD was that they were unable to compensate their investment managers in line with the value they created. The ability of the limited partnership structure to align compensation incentives meant that the best talent was flocking away from firms like ARD and toward the ones set up as limited partnerships, putting the final nail in ARD's coffin and ensuring that the limited partnership model would come to dominate venture investing for the decades that followed, right up to today.

Importantly, limited partnerships also offer significant tax benefits for both limited partners and general partners. In fact, as Nicholas points out in *VC: An American History*, their rise in popularity in the mid-twentieth century was driven in large part by a desire for investors to limit their tax liabilities. Limited partnerships allow LPs to claim a variety of deductions and to write off losses in ways not available on other types of investments. General partners, in turn, are able to take advantage of the controversial carried interest loophole, which allows them to report their share of the profits from the funds as long-term capital gains rather than income.

All these things—the aligned incentives through compensation, the tax benefits, the minimal oversight, and the limited liability—made limited partnerships a perfect vehicle for venture capital investing, and allowed the model to take off. Most of the pioneering venture capital firms that adopted it outperformed the public markets with their earliest funds, replicating the type of success ARD demonstrated—and entrenching belief in the power law—but with much lower overhead and appealing compensation and tax benefits to boot.

The successes that emerged from these early firms created momentum for more venture capital, which was by the late 1960s coming to be defined by its characteristic power-law-distributed fund profiles and its laissez-faire corporate structure. Today, those two elements are so intrinsic to what venture capital is, it is generally taken for granted that this style of investing is the only style that can achieve outsized returns in the field of high-tech investing.

VENTURE CAPITAL ENTERS THE MAINSTREAM

Despite the early glimmers of promise, venture capital remained a backwater of the American financial system until the 1970s. Then, the formation of a key congressional task force and the birth of an industry lobbying group paved the way for venture capital to enter the mainstream.

In 1976, the Small Business Administration convened the Task Force on Venture and Equity Capital for Small Business, colloquially known as the Casey Task Force after its chair, William J. Casey, an influential member of Republican presidential administrations who was once the chair of the SEC under Richard Nixon and would go on to lead Ronald Reagan's CIA (Central Intelligence Agency) in the 1980s. The Casey Task Force's stated purpose was to find a solution to the lack of abundant risk capital in the startup and small business ecosystem. Despite the work of Doriot and ARD, and the pioneering venture capital firms that followed in the 1960s, the conveners of the task force still believed that not enough investors were willing to put risk capital into new innovative business opportunities.

Recent research from M. R. Sauter, a professor at the University of Maryland, makes a compelling argument that the Casey Task Force may have overstated the lack of available capital for unproven high-technology companies in the 1970s in order to encourage policy interventions that would bolster their own existing investment models. Casey himself was a venture capital investor on the side, and the Casey Task Force's report was used as justification for a number of deregulatory and supply-side policy proposals throughout the 1980s. Ultimately, Sauter argues, the Casey Task Force report played a major role in validating and entrenching the power law and limited partnership approach to venture capital that has come to dominate startup investing.

The most consequential of the task force's recommendations was to allow pension funds to invest in venture capital funds. Pension funds were a hot policy topic in the 1970s. Most famously, the Teamsters union, run by Jimmy Hoffa, was embroiled in a scandal involving misuse of pension funds; and the stagflation era, when both interest rates and inflation were stubbornly high, caused many pension funds to fail to meet their obligations. In response, policy-makers were eager to protect employee pensions from unnecessarily risky investments. Congress passed the Employee Retirement Income Security Act (ERISA) in 1974 to do so. ERISA contained a clause, called the "prudent man rule," which required fund managers to act as fiduciaries "with care, skill, prudence and diligence" when managing funds. According to the financial sector, the prudent man rule was a major impediment to economic growth. In their view, it was overly subjective and vague and would create a chilling effect that incentivized fund managers to be too cautious. It would not only tamp down the growth potential of pension funds, they argued, but also dampen investment in high-growth and high-risk areas like the emergent high-technology industry.

As the Casey Task Force was doing its work, the nascent venture capital industry established a lobbying vehicle, the National Venture Capital Association (NVCA), to carry the torch for the recommendations that came out of the Casey report. The NVCA, working in close collaboration with the Casey Task Force, systematically went to work to weaken the prudent man standard. Trying and failing twice to loosen that standard through congressional action, probably because the fraught state of the economy at the time made it politically infeasible to weaken financial regulations, the groups instead focused on influencing the rulemaking process. In 1979, they succeeded in getting the Department of Labor (which oversaw the enforcement and implementation of ERISA) to clarify the prudent man rule in such a way that allowed pension funds to invest in venture capital. The effect was immediate and immense. In 1978, venture capital funds were managing $218 million in investor funds. By the end of 1979, the year the rule changes took effect, more than $2 billion had been invested, and that number rose to $3 billion by 1988, much of it from pension funds. Most experts agree that these changes to ERISA are the single most important factor contributing to venture capital's growth in the 1980s and 1990s.

The ERISA rule change wasn't the only adaptation of America's regulatory structure that allowed venture capital to become a cornerstone of the country's financial system. Buoyed by their success, the NVCA joined with other free marketeers to push hard to lower the top marginal capital gains tax rate from 50 percent to 28 percent, a fight they also won. This was the heyday of supply-side economics, when policymakers were convinced that by making it easier (and more lucrative) for business to operate, they would create economic growth that trickled down to everyone. The supply-side argument for lowering the capital gains rates was that taxing capital gains at higher levels disincentivized the kind of investment the country needed to spur economic growth. If investors could take home more of their gains, the argument went, they would be more likely to take those risks in the first place—and to pump profits back into the economy in the form of additional investment.

Experts disagree on whether the changes in capital gains tax rates had any impact on how much money venture capitalists invested in startups, and there isn't much data to prove that it did. What is clear forty-plus years later is that supply-side economics didn't result in better economic outcomes for the country as a whole. Inequality has skyrocketed and wages for working people have stagnated even in the face of increasing productivity. Despite this, the NVCA continues to fight hard to maintain low tax rates for investment income (the top capital gains tax rate has been lowered even further since the 1980s, to 20 percent today).

Today, venture capital is squarely at the center of the world's economy. No longer a niche investing vehicle for a few eccentrics, venture capital is practiced by a wide array of actors from celebrities like Serena Williams, Ashton Kutcher, and Katy Perry to sovereign nations like the Kingdom of Saudi Arabia. According to the NVCA, which has cemented itself as a major lobbying force in Washington since its early days advocating for ERISA reform, more than $128 billion was raised by venture capital funds in 2021 and $330 billion in venture capital dollars were invested in startups.

By 2012, when I arrived in San Francisco, Silicon Valley was the envy of the world. Countries from South America to Asia were trying to replicate its success. As central banks lowered interest rates to zero in the 2010s, limited partners had almost unlimited cash and were desperate to put it somewhere that could deliver outsized returns. Entrepreneurs—and some VCs—were becoming celebrities, gracing the cover of magazines from *Forbes* to *Vanity Fair*. Shows like *Shark Tank* brought the concepts of equity investing to the masses on broadcast television. Instead of going to work on Wall Street, graduates of the top business schools were now flocking to California to pan for startup gold.

As the venture capital system matured and expanded, what was once an observation hardened into an incontrovertible law of physics: when venture capital funds are successful, they are power law distributed. By the 2010s, venture capitalists weren't just of the belief that power law returns occurred *as a natural result* of investing in high-risk startups. They believed that power law returns *could be reverse engineered* by optimizing their funds to contain as many potentially astronomically returning companies as possible. Today, the power law has now become more than definitional; it is a methodology pursued by almost every venture capitalist in business.

If you listen to VCs talk about the power law, this distinction becomes clear. It's what Bill Gurley meant when he said that venture capital was a grand slam business. Chip Hazard, a veteran VC with over twenty-five years of experience at some of the top firms, has said: "Great venture capitalists and venture capital funds are defined by massive winners. The best investors fully internalize this power law of returns dynamic and seek to back companies that can become outliers—massive wins are all that matters in driving returns."

For venture capitalists, the job has become not about finding companies that are creating breakthrough innovations that could spur large-scale value creation, but about pursuing power law returns that can make them very rich.

By nature, this pursuit of power law returns needs a lot of failures. After all, a power law is defined not just by its huge upside but also by its long tail of very small values. In his book *The Power Law*, Sebastian Mallaby describes the returns of Horsley Bridge, one of the most prestigious limited partners in the venture capital game. Between 1985 and 2014, Horsley Bridge invested in venture capital funds that, altogether, invested in 7,000 startups. Of those companies, just 6 percent of them accounted for 60 percent of the total returns. And when you look deeper, at the individual funds, those that returned more than five times the original investment to its limited partners (what would be considered a very successful fund) actually had more losers—investments that returned less than the original capital—than the funds that returned two to three times the original investment.

The conclusion that some venture capitalists have drawn from this data is that to build a successful fund, you not only have to optimize for big winners, you also have to optimize for a significant number of failures as well. Not that venture capitalists want their portfolio companies to fail, but rather, they see failure as an indicator that they are taking on the level of risk that also produces the outsized successes—that their attempts to reverse engineer power law returns is working. According to Benedict Evans, a former partner at Andreessen Horowitz, venture capitalists "invest in a particular type of startup, one that is much more likely to go to zero, but which has the potential, if it does succeed, to produce something very big." In other words, the aim of venture capitalists is not to identify the companies that are building something of great value and, if successful, would have huge financial upside. The aim of venture capital is to identify the companies that have huge financial upside, period. What those companies are building, what value they are creating in the world, is of secondary importance.

A theme that emerged from my dozens of interviews with entrepreneurs and investors alike is that rather than simply seeking out inherently risky business opportunities, much of what venture capitalists do is *create* risk for their portfolio companies. Since investors don't know which of their portfolio companies will be the big returners for the portfolio, they hedge their bets by pushing all of them to apply the same get-big-fast playbook. As Nick Grossman, a partner at Union Square Ventures, told me, "Most funds underwrite each investment assuming it can be a fund returner." By pushing the companies in their portfolios to achieve the massive scale it would take to drive a power law return, they enable unnecessary risk-taking.

Venture capitalists don't just endorse the power law theory as key to success, they also believe and espouse that taking any other approach to venture capital investing *can't* work. That investing in early-stage tech startups is definitionally about pursuing power law returns. Jason Calacanis, who found success investing in companies like Robinhood and Uber and has since become a venture capital influencer, extended Bill Gurley's famous baseball analogy, saying "it's just not possible to make the single and double concept work [in venture capital]. Singles and doubles is what public market investors do or late stage investors do." Almost all the dozens of venture capitalists I spoke to for this book expressed generally the same sentiment: venture capital investing is about power laws, and if you're not trying to generate them, then what you're doing isn't venture capital. This was such a fundamental belief that when I asked many of them to examine the assumptions embedded in those statements, they responded with the kind of impatience and exasperation that parents express when kids ask why the sky is blue. It just is!

So, what's the problem with this mindset? Aside from the increased risk it creates for its portfolio companies, venture capital's commitment to achieving power law returns—and VCs' blanket dismissal of any other approach to venture capital investing—is the single greatest driver of the negative societal outcomes created by the tech industry.

This influence is felt first in the selection process, where venture capitalists decide which ideas will get pursued and which won't. If founders can't tell a story in their pitch about how their company can achieve venture-scale returns—whether that story is realistic or not—they fail before they even get out of the starting gate. The yarn that founders are asked to spin when pitching VCs has some very standard elements. They must be able to convince the funders that their total addressable market, or TAM, is large enough to achieve venture scale, and that the team they are building to meet that market demand is capable of delivery. As one investor put it succinctly when I asked him what he looks for when evaluating a pitch, "team, team, TAM and team."

The founders, for their part, are often eager to play along. We'll hear stories from some of these entrepreneurs in later chapters. Many of them spoke of either not understanding that there were other options to fund startups besides venture capital or being in thrall to the idea of running a venture-backed company. Or they were resigned to the fact that venture was the most likely pathway to helping them build the world-changing company they wanted to build. Almost all of them told me how they had been subtly and implicitly asked to shape the story of their company, over the course of dozens or even hundreds of pitches, to make it fit a venture mold, regardless of whether that story was plausible. "[Venture capitalists] asked us to lie to them in very specific ways," one of the founders told me.

There is a fundamental paradox at the heart of this process: the vast majority of venture capital investments aren't capable, for whatever reason, of generating venture-scale returns. But VCs will attempt to fit all their investments into that box anyway in order to find the few that will.

VCs are aware of this paradox. It isn't uncommon to hear a VC say unequivocally that venture capital is not the right form of capital for most startups. "I sell jet fuel, and some people don't want to build a jet," Josh Kopelman of First Round Capital once said. And yet, if you look at their investments, venture capitalists are all too eager to put money into these non-rocket-ship companies anyway. In many respects, they don't have a choice. They have more money from limited partners than the universe of truly venture-scale companies really need. Besides, without a lot of low returners in their portfolio, how will they get the power law curve they desire? So, they shoehorn companies into their portfolios, knowing that venture capital is probably bad for many of them. Then they push the companies to get as big as possible as fast as possible, damn the consequences.

"SILICON VALLEY'S FAVORITE GROWTH STRATEGY"

Once the theater of the pitching process has taken place and investments have been made, venture capitalists now get to work enforcing the methods that they believe will create the best chance for their portfolio companies to hit huge returns. There is perhaps no better articulation of this approach than *Blitzscaling* by Reid Hoffman and Chris Yeh, published in 2018 at the height of the last tech boom cycle. Hoffman is a prolific and well-respected entrepreneur and investor. Once part of the original team at PayPal, he parlayed his success to found LinkedIn, which he sold to Microsoft in 2016 for $26 billion. He has been involved in venture capital as a partner at Greylock Partners since 2009. *Blitzscaling* is his advice for entrepreneurs who are, as he says, trying to take their companies from startup to "scale-up."

Everything you need to know about *Blitzscaling*, which renowned tech publisher and investor Tim O'Reilly has called "Silicon Valley's favorite growth strategy," is in its title: this is a how-to guide for getting big, fast. It professes to be relevant for any kind of company, venture-backed or not, but is clearly informed by Hoffman's experience with venture capital. Hoffman and Yeh instruct entrepreneurs to inefficiently flood the zone so that their companies can amass enough of the market to dominate. Much as with the Nazi military strategy that gives the book its title, the goal is to overwhelm and engulf the competition before they have a chance to gain a foothold. Blitzscaling is "an accelerant that allows your company to grow at a furious pace that knocks the competition out of the water," according to the authors. The audience is "anyone who wants to understand the techniques that allow a business to grow from zero to a multibillion-dollar market leader in a handful of years."

The way to do this, according to Hoffman and Yeh, is to prioritize speed over efficiency. Use a sledgehammer, not a scalpel. Collateral damage is a feature, not a bug. This kind of approach demonstrates the kind of irrational and unnecessary risk that venture capitalists encourage companies to take on in order to deliver their returns.

Beyond the risk to the businesses themselves, this methodology asks the companies to operate with a certain amount of intentional thoughtlessness—creating unnecessary risk for the rest of us in the process. Hoffman and Yeh instruct their students (the book was born out of a class they taught at Stanford) to "embrace counterintuitive rules like hiring 'good enough' people, launching flawed and imperfect products, letting fires burn, and ignoring angry customers."

According to the *Blitzscaling* method, companies are forced to make every decision with hyper profit maximization in mind. There is no room to leave money on the table in order to take other things into consideration. If the choice is between doing something that will increase a company's valuation and doing something that may be better for society, startups will be hard-pressed to choose the latter path. This dynamic doesn't just incentivize bad behavior, it requires it. At these levels, choosing to leave money on the table is akin to choosing death. To avoid that fate, Blitzscaling startups push the risks onto the rest of us.

Hoffman and Yeh hold up Uber as a prime example of how this

flood-the-zone strategy can be deployed effectively, and rely heavily on anecdotes from the company's ascent to make their point about how they think other startups should approach growth. I am reluctant to use Uber as the example for how venture-fueled excess can cause messes that the rest of us are left to clean up, because its story is so outlandish it makes it easy to discount as an outlier. But Uber's journey is such a perfect object lesson that it is worth recounting at least some of the details here.

Uber got its start in 2009 but didn't enter its Blitzscaling phase until 2013 when it expanded to almost 100 countries around the world and grew from hundreds of employees to 15,000 in just a few years. By 2017, Uber had raised $15 billion in venture capital, which it used to overwhelm the market and its competitors. "Uber often uses heavy subsidies on both sides of the marketplace when it launches in a new city, lowering fares to attract riders and boosting payments to attract drivers," explain Hoffman and Yeh. "By paying out more than it takes in on those early trips, Uber is able to reach critical scale faster than a more conservative competitor."

Uber's strategy relies on more than just subsidizing the cost of rides. Its Blitzscaling approach drove it to steamroll into new markets, operating without the required permits to do so and daring local officials to make it stop. The overwhelming imperative to expand its market share led executives to begin operating the company with a siege mentality that justified any behavior to counteract what they saw as their opposing forces. In *Super Pumped*, Mike Isaac's detailed history of Uber's evolution, Uber's founder and then-CEO, Travis Kalanick, is described as being "engaged in a crusade" against the powers that be "who were colluding to keep taxi service bad and overpriced." In 2017 in reporting for *The New York Times*, Isaac revealed the existence of a program at Uber called Greyball that identified regulators in unfriendly cities who had downloaded the Uber app to gather evidence of its operations. Greyball served those regulators a dummy version of the app when they tried to hail a ride in order to throw them off the scent and maintain its illegal operation.

Uber was focused on ensuring that its main rivals in the app-based rideshare market couldn't gain a foothold. Blitzscaling encourages its proponents to take an extreme us-versus-them stance toward competition, which Uber readily demonstrated in its behavior toward Lyft. Uber employees were instructed to hail Lyfts and, when their cars arrived, cajole the drivers to switch to Uber. The company also enlisted employees to book, and then promptly cancel, thousands of Lyft rides in order to inject chaos into the competitor's operations.

In 2022, a whistleblower (now known to be an Uber lobbyist named Mark MacGann) leaked thousands of internal documents exposing the extent of Uber's shady dealings as it sought to expand around the world. "We're just fucking illegal," one Uber executive wrote to a colleague. As the company was facing investigations in multiple countries, Uber built a kill switch into their servers, which they activated when police conducted raids on their offices. In an attempt to gain access to the Moscow market, Uber showed favor to oligarchs close to Vladimir Putin, many of whom have been living under sanctions since Russia's invasion of Ukraine. And the company repeatedly sought to prevent scrutiny of its tax avoidance schemes by pointing tax collectors to drivers instead. All of this, presumably, was justified by the company's being able to amass the market share it needed quickly enough to drive taxi companies out of business.

The internal culture that the Blitzscaling methodology engendered at Uber was, unsurprisingly, exceedingly toxic. Around the time Project Greyball was uncovered, a former Uber software engineer named Susan Fowler published a searing report of her experience working at the company. She recounted rampant sexual harassment, systematic misconduct on the part of executives, and many other instances of unethical, threatening, and sometimes illegal behavior. At one point, Uber executives floated the idea of spying on and releasing sensitive personal information about a female journalist who had been critical of the company. The company also allegedly tried to sweep under the rug the alarming rise of reports of sexual assaults by drivers on female passengers.

Fowler's revelations set off a reckoning that resulted in several lawsuits, fines from regulators, and ultimately, Travis Kalanick's departure as CEO.

The company went on to build perhaps the worst reputation of any venture-backed company still in existence, and has inflicted serious harm since its founding. Maybe Uber would have always gone down this path, whatever their funding model was. But it's worth asking how much of this damage could have been avoided if the company and its executives weren't so desperate to Blitzscale, and if venture capitalists hadn't fed billions of dollars into the furnace that enabled it to do so.

Hoffman and Yeh, for their part, do state that they think when entrepreneurs Blitzscale, they should do so "thoughtfully and responsibly," although that doesn't quite square with a methodology named for a Nazi military strategy fueled by the German army's prolific use of amphetamines.

———

The macroeconomic environment has changed a lot since Uber launched its Blitzscaling campaign, making it much less appealing for venture capitalists to support this kind of strategy. As the Fed began raising interest rates in 2022 in response to stubborn inflation, the venture capital money spigot got turned way down. There was no longer enough free cash to pump into Blitzscaling businesses, and a lot more scrutiny on business models that required, as Blitzscaling does, "prioritizing speed over efficiency." Now, instead of extolling scale, investors began trumpeting the virtues of business fundamentals and focusing on revenue over growth. Many startups were caught scrambling, trying to pivot their companies to create something that customers actually want to buy at the price it actually costs to deliver it.

This new reality has probably killed the Blitzscaling phenomenon for now, but not before it made a real impact in Silicon Valley. Hoffman and Yeh may not have meant for readers to take it as a guide that any entrepreneur could follow, but that was the effect. Blitzscaling has become synonymous with successful growth and many hundreds of startup founders have infused its principles into their strategies. In many ways, all *Blitzscaling* did was to say the quiet part out loud: that successful startups, those that will be the winners in venture capital funds delivering power law returns, grow fast with little regard for the havoc they might wreak on the rest of us.

THE RISE OF THE MEGAFUND

As low interest rates persisted well past the end of the recession and into the 2010s, the venture capital industry exploded. The desperate eagerness of the limited partners to put their piles of cash to work in high-returning asset classes was palpable. There were more venture capital firms raising money for more funds than ever before. From 2007 through 2023, the number of venture capital firms increased almost fourfold, from 947 to 3,417. It wasn't just the number of venture capital funds that was increasing astronomically. The amount of money invested by LPs in American venture capital funds went up more than 500 percent between 2007 and 2022, before the falloff driven by higher interest rates. This phenomenon of increasing fund sizes gave rise to the so-called megafund, that which is larger than $500 million. In 2018, megafunds raised 44 percent of all the money allocated to venture capital funds. By 2022, that proportion had risen to 70 percent.

Fund size, more than any other single feature, says the most about how likely the companies within the fund will be pushed hard to achieve Blitzscale-like outcomes. This is because of the math that goes into calculating overall fund returns: the larger a fund gets, the bigger the winners in those funds need to be to enable venture capitalists to deliver the returns that limited partners expect. A successful venture capital fund typically expects to return three to five times the original amount invested; that means megafunds (depending on their stake in each company) need to contain at least one multibillion-dollar winner—and that's just for the smaller megafunds. The megafunds with over a billion in value must produce even more big winners simply to break even.

To illustrate how this works, a former venture capitalist named Evan Armstrong did the math on one megafund in a detailed essay published in 2022. Index Ventures' $3.1 billion fund raised in 2021 has a stake in Figma, a company that makes a very popular suite of graphic design collaboration tools. In 2022, Figma announced it was being acquired by Adobe for $20 billion. While regulators eventually quashed the deal because of antitrust concerns, had it gone through, it would have been one of the most successful acquisitions in Silicon Valley history.

Index, as Armstrong points out, owned a 13 percent stake in Figma, which would have valued its stake in the startup at about $2.6 billion. We don't know exactly how much Index put into Figma over the years, but estimates say the firm would have made somewhere between a 30x and 90x return. By anyone's measure, that is a wildly successful investment return. But as Armstrong points out, it wouldn't be enough to match the size of Index's fund—the criterion venture capitalists typically use to judge whether a single investment reaches grand slam territory. In order for Index to deliver the three to five times return LPs expect for the overall fund, the rest of the companies in that fund would need to collectively earn a valuation of at least twice what Figma had earned.

To grasp how difficult that is, it's important to understand how few companies have the ability to reach those heights. As Armstrong points out in his essay, the math for these funds means that investors are probably setting the grand slam bar at $50 billion—yet, as of this writing, there are fewer than fifty public tech companies that hold this status, and just a handful more that are still privately owned.

The pressure this puts on the companies within a megafund's portfolio is enormous. They all absolutely must push as hard as they can to achieve maximum market share and valuation. There is no leeway to leave a growth opportunity on the table, no matter what the consequences may be. The speed and urgency with which these companies must move increases the carelessness in their operations. As we'll see throughout this book, that pressure creates fallout across the economy while the venture capitalists and limited partners who profit from these dynamics mostly bear no cost.

Armstrong isn't the only one who has made the point about the challenges facing large funds. In a 2017 study, researchers found that the larger a fund gets, the worse it performs compared to smaller funds. As Harvard Business School professors Josh Lerner and Victoria Ivashina say about the study in their book, *Patient Capital*, "Groups that do not increase their size continue to perform at the same level (in fact, in the case of venture capital they improve their performance). But those whose fund size increases sharply experience sharply lower returns: a doubling of fund size translates into a reduction of [internal rate of return] by roughly four percentage points."

By now you might be wondering: Why do venture capitalists bother raising funds this large if they are almost guaranteed not to meet expectations? The answer, in large part, lies in the compensation structure of venture capital. As I described earlier, venture capitalists are paid through two streams: they take a (typically) 20 percent cut of the profits the fund earns, and they also are paid an annual management fee that is usually around 2 percent of the total value of the fund. The bigger the fund, the more money they can bring in through the management fee. Since the cost to manage a fund doesn't increase in proportion to its size, the general partners in the VC firm get to personally pocket more money in fees the bigger the fund gets *without having to deliver any results*. That's a pretty good deal if you can get it, and a very strong incentive to try and raise as much money as possible, regardless of how the fund performs.

This begs an additional question: Why do the limited partners put up with this? If they're less likely to have their return expectations met and they know the fund managers are siphoning off huge fees regardless of performance, why not divert money to smaller funds? The answer to this question is multifaceted and much more complex. But, as we'll discuss in more detail in chapter 6, it is the key to fixing the venture capital system. For now, suffice to say that the market dynamics of so much money chasing so few truly venture-scale opportunities creates a sense of FOMO that drives many limited partners to grudgingly accept the extortionate fee structure. Even when limited partners do try to push back on astronomical fees, they don't have much choice but to continue making commitments to the funds they thought could secure a piece of the next multibillion-dollar startup. The LPs are selling themselves to VCs as much as VCs are selling themselves to the LPs. If they develop a reputation for being difficult, they can be frozen out of the best investing opportunities for decades.

There may be signs that megafunds are falling out of favor. In the era of high interest rates, the amount of money allocated to megafunds has dipped dramatically. High-profile firms like Insight Partners and Tiger Global have conspicuously reduced the size of their latest funds (though they still are in the multibillions of dollars). But, as interest rates begin to moderate and the growth of the AI sector increases the sense of FOMO among limited partners, there is nothing stopping fund sizes from ballooning once again.

THE VALUE OF A SENSE OF URGENCY

Alongside employing the limited partnership structure, the early venture capital firms made one other critical decision that has come to define venture's impact on the economy: they limited the lifespan of their funds to an arbitrary period of five to ten years. While the tax advantages and limited liability provided by the limited partnership model were enticing to limited partners, there was still the reality that these investments would remain illiquid for an indeterminate length of time. By creating limited partnerships that simply dissolved after a fixed period, LPs.at least had the assurance that they would receive whatever returns they were owed within a specified time frame. The time-limited nature of these funds feeds into an up-or-out mentality that creates an additional incentive to push portfolio companies hard to reach a liquidity event on a relatively short time schedule.

While ten years may seem like a long time to hold an illiquid asset, in practice, portfolio companies have only a couple of years to prove their value to investors. We'll talk more about the funding cycle of venture capital, and how it reinforces the incentives that drive harmful outcomes, in chapter 5. For now, it's enough to know that while fund lengths are officially about ten years, venture capitalists typically start raising new funds every two to three years. This, of course, creates an opportunity to bring in even more fees from limited partners. But to convince limited partners to support yet another fund, the VCs must have some metrics to show that their current investments are on track to be grand slam successes. While venture-backed startups may have about ten years to reach a definitive liquidity point, they have much less time to show they can achieve the market dominance required by venture investors.

Short time horizons also mean that investors—both limited partners and venture capitalists—don't have an incentive to care very much about the long-term health of the companies in which they are investing. This creates a moral hazard whereby the earliest investors are incentivized only to achieve the highest possible valuation at a specific point in time—the liquidity event—rather than set the company up to create sustainable value in the long run. Others—public market investors but also the general public—are left to deal with the fallout when the decisions that benefited early investors turn out to have negative consequences further down the road. As Warren Buffett once famously said, quoting the investor Benjamin Graham, "In the short run, the stock market is a voting machine. Yet, in the long run, it is a weighing machine." By this he meant that the value of a publicly traded company in its early days post-IPO is driven by popularity more than by true business fundamentals. Over time, as the hype wears off, stock prices reflect the real value of a company as public market investors have a chance to judge it on the merits. But venture investors are usually only required to hold their stake in companies for three to six months after an IPO—hardly enough time to judge the long-term value of a company. That early investors can cash out before a company's true valuation has been fully weighed by public market investors, and bear none of the consequences for any long-term ill effects of what it took to reach those heights, is a remarkable moral failure of our capitalist system.

Short-termism in venture capital also limits the types of problems venture capitalists are willing to address. As Lerner and Ivashina point out in *Patient Capital*, the areas where we most need venture capital to fund breakthrough innovations—like clean energy and infrastructure—are often areas that require much longer time horizons than venture capital currently allows.

In 2007, John Doerr, who has been a partner at Kleiner Perkins for over forty years and is one of the most respected venture capitalists in Silicon Valley, gave a TED Talk in which he made the case, through tears, that climate change was the biggest challenge facing humanity. For investors, Doerr argued, there was an opportunity—bigger than the internet—not only to make a lot of money but to save the planet from impending doom. Doerr was committing $100 million from his latest fund (a lot by the standards of venture capital fund sizes in 2007) to clean tech solutions, and he called on the investors in the audience to follow his lead. They listened. Venture investments in clean tech surged between 2007 and 2012 as VCs decided that clean tech was, in fact, the next big thing.

By 2013, however, investors were getting impatient. Eager to be out raising their next fund—and raking in the fees that would come along with it—VCs were learning that the process of commercializing clean energy solutions was much more complicated and capital intensive than the software projects they were used to. The science was harder, the politics were thornier, and the cost to reach scale was much higher. As the earlier clean tech investors began to assess their portfolios five years in, they realized that their portfolio companies were not close to having venture-scale exits. Less than 2 percent of clean energy companies that received investment between 1995 and 2007 had gone public.

This spooked fund managers. By 2015, investments into clean tech companies had dropped to under $1.5 billion, down from a peak of almost $4.5 billion in 2011. While the tough political and scientific realities intrinsic to the clean tech space were mostly out of the control of investors, John Doerr caught a lot of flak from the VC community for being "wrong" about the clean tech opportunity. While it isn't clear that his commitment to clean energy investing had anything to do with it, in 2016, he stepped out of his role as an active investing partner at Kleiner Perkins.

If investors had had a longer time horizon, this story might have had a different ending. Some of those companies may have been able to catch the green energy wave that was coming in just a few short years. I'm not the first to suggest that longer time frames for venture funds might be necessary and even beneficial for investments in capital intensive, deep science areas like clean tech. The lack of this patience has meant that the most promising clean tech solutions of the last decade had a much harder time finding the capital they needed to develop their products, potentially slowing our progress on climate change.

The decisions the earliest venture capitalists made about how to structure their funds, the conclusions that subsequent investors drew from their experience with respect to power law returns, and the increased intensity created by the ballooning size of venture capital funds together have distorted venture capital away from its original mission to fund innovative but risky technological breakthroughs that move the ball forward for the economy and society. Where venture capitalists used to play only on the margins of the financial system—a natural result of the fact that their aim was to fund breakthrough technologies rather than to engineer multibillion-dollar exits—VC is now squarely at the center, infused into the most mundane corners of the economy.

That world-eating mentality—which has not only pushed venture capital into areas it has no business being but also has taken its focus off the solutions to some of the biggest challenges society faces—is resulting in dangerous harms for the rest of us.

2

THE FOUNDERS

HOW VC UNDERMINES ENTREPRENEURSHIP

I didn't mean for this book to be one about entrepreneurs. But as I talked to more and more people who had raised venture capital to start what they hoped would be world-changing companies, it became clear that their experiences, more than any broad analysis I could do, encapsulate the damage that venture capital inflicts. The choices they are forced to make, the expectations that are placed on them, the despair they often feel in the face of what seem like no-win decisions, all reflect the way venture capital is distorting our economy for the worse.

There is a lot of variation in both the type of people who found startups and the experiences they have raising capital. Over the course of my research, I have heard dozens of stories from entrepreneurs about what motivated them to start a company, how they approached raising capital to get their idea off the ground, and what happened when they started working with venture capitalists. I also read the accounts, some firsthand and some works of journalism, of many other founders. I found that their stories reflected certain patterns and that they tended to sort into four broad categories.

1. *The Maximalists*: These founders are fully on board with the values of venture capital and feel no compunction about running the Blitzscaling playbook, no matter the consequences. They seem comfortable with the charge given to them to grow as big as possible as fast as possible, and mostly seem to be motivated by glory and riches. They adopt a venture capital worldview and are denizens of the "fake it 'til you make it" church of American capitalism. They have been well rewarded by the venture capital community for their ability to spin the best tale and do whatever it takes to make the vision a reality. A few of them have faced criminal charges for this. Many others, somehow, managed to stay in the well-trodden gray area in American capitalism between audacious enterprise and outright fraud. A lot have been rewarded for their bad behavior. There are fewer Maximalists than you might imagine, but they are usually the ones you hear about; when things go wrong, they tend to go spectacularly wrong. The Maximalists are the poster children for what people perceive as the rot in Silicon Valley.

2. *The Aligned Capitalists*: This set of founders don't have the same lightly sociopathic tendencies as the Maximalists, but they are generally very comfortable with the VC mindset. They have a

zeal for entrepreneurship. The idea of growing a business to massive scale impacting millions of lives animates them much more than the specific mission of the business they are building. Hence, they tend to go along with venture capital demands without much fuss or hand-wringing. They take a "they know best" approach to what their investors tell them and they're very happy to bask in the prestige that being funded by brand-name firms brings to them and their companies.

3. *The Conflicted Altruists*: These entrepreneurs are, by far, the most interesting to me, and are the ones with whom I spent the most time. They are driven by their desire to solve a specific problem and believe that growing a scalable business is the most effective way to address it. They are committed to capitalism, but they also believe that businesses have responsibilities beyond creating the largest possible returns for their investors. For them, it is okay—in fact, often necessary—to leave money on the table in order to optimize for outcomes that create other kinds of value. They are often somewhat reluctant to take venture capital but tend to view it as a necessary evil. They face a Faustian choice: take the money and hope they can get off the venture capital treadmill before it consumes them, or give up the idea of building a business at all. Once on the VC track, these founders are often forced to make painful, even traumatic, decisions that leave them disillusioned about the startup ecosystem.

4. *The Counterculturalists*: The last group are the ones who opt out of traditional VC altogether and try to find another way. Many of them used to be Conflicted Altruists or even Aligned Capitalists, but terrible personal experiences with past startups radicalized

them away from the world of VC altogether. They have made their peace with building companies that may be smaller than they might with VC investment, or they are savvy enough to understand from the beginning that the companies they are trying to build may actually be more successful with different forms of capital.

What all four types have in common is that they, consciously or subconsciously, orient themselves in relation to venture capital. As venture capital has increasingly become a monoculture in which most firms are trying to reverse engineer power law returns, entrepreneurs have internalized that method and organized themselves around it. Some are eager, excited, and willing to play the game. Others find themselves fundamentally opposed. A larger group in the middle concede to reality and go along because they don't think they have much choice. But none of them escape the influence VC has on the world of entrepreneurship. Along the way, this dynamic has distorted what it means to start a successful business in the twenty-first century.

Alicia Rouault, the cofounder of LocalData who you met in the introduction, was a Conflicted Altruist through and through. Her experience of naively walking into venture capital because she didn't feel she had any other options wasn't an accident. Entrepreneurs like Rouault might develop this belief about the supremacy of venture capital because they hear it most explicitly from the advisers and seed investors who help them get their ideas off the ground. Startup incubators and accelerators sprang up in abundance during the late 2000s and early 2010s, performing a critical role in the venture-backed startup ecosystem. Accelerators gather founders to-gether into time-limited cohorts, providing them with capacity and resources (including, usually, an equity investment in the company) alongside access to mentors and peer-learning opportunities that help take startups to their next stage of development. Accelerators are a primary conduit for these founders to connect with investors, and the accelerator mentors have an enormous amount of influence over how founders think about structuring their business models and pursuing capital investment. Most accelerators end with a pitch day, when the participants give a presentation to a group of investors hoping to rope some of them in to invest in their companies.

By far the most prestigious of these accelerators is Y Combinator. YC, as it is known, was founded in 2005 by Paul Graham and Jessica Livingston after Graham sold a company he founded to Yahoo! Since then, over 4,000 startups have gone through YC's three-month boot camp, including more than a handful of name-brand Silicon Valley companies like Reddit, Airbnb, Dropbox, Coinbase, Stripe, and DoorDash. Companies that are selected into YC cohorts are given $125,000 in exchange for a 7 percent equity stake in the com-pany. But far more important than the money, YC founders are con-nected with mentors who help them learn how to think like a venture-scale startup. As YC's former president Geoff Ralston told *Wired* magazine in 2021, "We're sort of like Crispr for startups. Startups come into YC with raw DNA. We edit the DNA so that they have the alleles that make it more likely for them to be successful." A key part of YC's gene-splicing curriculum is exposing its cohorts to successful investors and founders who can deliver the message in an authentic and credible way—all while making the new founders feel like part of an exclusive community. There are guest lectures, salon dinners, and informal happy hours that serve to cement the power law belief system. Once founders graduate from YC, they have access to a private virtual space called Bookface where all YC founders past and present can connect and share advice. This environment creates a powerful set of cultural norms that define what success looks like in Silicon Valley: fast-growing startups that go on to become what are called unicorns, startups that are valued at more than $1 billion.

Because of the track record YC has developed, it has become a model for the cottage industry of other Silicon Valley accelerators that came into existence in the mid-2010s. Their methodology loomed large over Silicon Valley at the same time that LocalData was getting off the ground.

Matter.vc, one of the startup accelerators in which the LocalData team participated, was run by Corey Ford, a media entrepreneur who has a background in both journalism and tech. Ford launched Matter in 2012 as a way to catalyze innovation in the news business at a time when media companies were frantic about the impact the internet was having on journalism's business model. News media startups that were trying to address the problem were struggling to find investors who understood their market, which was mostly too small to be of interest to traditional venture capitalists. "I saw a lot of companies doing good work, solving big problems but they were never going to be grand slams," he told me. It's fair to say LocalData fell into that category.

Ford recognized that entrepreneurs like Rouault were facing societal headwinds that, subtly and not so subtly, shaped the way they thought about the viability of their businesses. "Entrepreneurs conflated fundraising with success and scale with impact," Ford said. "They confused venture capital funding with a signal that they're doing the right business or not."

For his part, Ford doesn't remember explicitly pushing LocalData to pursue traditional venture capital funding. He says his approach at Matter was to help founders decide for themselves what kind of capital was the right fit for their business. But Ford concedes that his success as an investor depends on the businesses he invests in finding future investors who will help them grow.

Early-stage investors like Ford are responding to an important feature of venture capital. Companies are funded in stages, with the expectation that as they hit certain milestones and prove with increasing certainty that the business can work (this is called de-risking), they will get larger infusions of cash from bigger investors. These stages are named alphabetically—Series A, Series B, and so on—and reflect how mature a company has become. Beyond Series D and E, the expectation is usually that a company will go public or get acquired and, by doing so, deliver returns to its investors. Seed-stage investing emerged more recently to invest in companies even earlier in their life cycles prior to a Series A. Now, there are even VCs that specialize in pre-seed investments that help entrepreneurs develop the ideas that go on to become investable companies.

Each subsequent funding round creates a value for the company, providing earlier-stage investors with concrete data showing how much their stake is worth. If their portfolio companies have hit the milestones that the next-round funders expect, they will typically place a higher valuation on the company than it had at its last funding round. At that point, earlier-stage investors are able to go back to their limited partners and tell them for the first time how well their investments are doing. Since the investment stakes are highly illiquid and since the companies don't yet have much of a track record, there is no other way to determine a startup's value than by how much a future-round investor is willing to put into it. It remains

this way until companies reach what's called a "liquidity event"—
they are either purchased by another company or they go public. It
usually takes a startup a decade or so to reach that point. In the
meantime, venture capitalists need something to show their limited
partners so that those limited partners will keep investing with
them, and so venture capitalists can collect the fees that come from
that stream of capital.

This dynamic highly incentivizes earlier-stage investors to mold
their portfolio companies into what they think later-stage investors
will find appealing. This isn't always what is in the best interests of
the companies, the people who are running them, or the rest of us.
These early-stage investors also need their portfolio companies to
actually *need* new rounds of investing. A company that reaches sustainability too soon because it has a business model that is producing
all the revenue it needs to grow, and doesn't require new capital investment to fund its development, is a bad thing for traditional venture capital investors because there's no mechanism for putting a
higher price tag on the company.

What does this all mean for investors like Corey Ford at Matter,
who are trying to spur innovation in a field where most of the good
business ideas that solve real problems probably aren't going to
reach venture scale? He didn't have many options. He'd prefer that
there was a more diverse ecosystem of later-stage funders whose
investment strategies aligned with the growth prospects of the
kinds of companies he was supporting at Matter. But those kinds
of investors—the kind that see the opportunity in a company like
LocalData—are few and far between. "What we needed was a doubles path rather than a grand slams path," he told me. "And the reality is that that path is not very clear and hardly exists. How do I get

[our companies] to the next phase? Where are the people who are investing in doubles?"

Those questions were ultimately existential for Ford: he had to shut Matter down in 2019 because his commitment to funding companies that were hitting doubles instead of grand slams meant he was unable to raise more money from limited partners.

Mainstream venture capitalists, for their part, understand that their success—their very existence—depends on the rest of the investor ecosystem, so they put enormous pressure on their portfolio companies to operate in a way that sends signals to future-round investors that the company can achieve venture scale. "In the old days companies competed on the merits of their products," Tim O'Reilly, the veteran Silicon Valley publisher and investor, told me. "Now they compete based on how much money they can raise. Building startups became a game of convincing investors instead of convincing the market."

***

In Silicon Valley culture, starting a business that dies trying to achieve behemoth status with exponential growth is a far more respectable outcome than simply building a small but sustainable company that grows linearly. These sustainable, slower-growing companies are known pejoratively as "lifestyle businesses" because they offer founders the opportunity, so they say, to have a higher quality of life than what Silicon Valley expects for founders of venture-backed startups. That is: working twenty hours a day, eating ramen noodles, and sleeping under their desk to save money on rent. The send-up of startup lifestyles as portrayed on the television show

Silicon Valley is remarkably close to real life. But the dripping disdain for lifestyle businesses that permeates Silicon Valley culture disguises deep-seated insecurities among venture capitalists. The truth is that lifestyle businesses represent a threat to the venture capital business model. If a company can become self-sustaining, it doesn't need to raise a future round of venture investment. If it doesn't raise a future round of funding, earlier-round investors don't get the data they need showing a higher valuation to justify their existence to limited partners.

LocalData probably could have survived as a lifestyle business, and figured out a way to keep the company growing without much external investment. Ironically, within a few years of its shuttering, the real estate industry got on board with technology and the market for what LocalData had been trying to sell finally appeared. There could actually have been a path to something approaching venture scale for LocalData if they had been around long enough to take advantage of the shift in the real estate sector. But Rouault was burned out and disillusioned and she was ready to move on. She and her partners decided the best way to wind LocalData down was to sell its core technology to a data analytics and software firm based in Detroit. That company is still operating as of this writing, a small glimmer of a happy ending for Rouault but far from the initial promise that LocalData showed.

Today, Rouault is contemplative about what the experience meant for her. "If I had to do this again, I would have stayed truer to our user-centered approach and been more comfortable with slower growth," she said, not just because of what it could have meant for the business but because of what it could have meant for cities like Detroit. "It was demoralizing to know that you were making something that helps real people, Black and brown communities in really poor neighborhoods in America, places where tech doesn't go," she reflected. "And then you're trying to raise capital for this thing that's succeeding, being told no, and you look over your shoulder and a grilled cheese company is able to raise the capital. The feeling was 'I guess we don't belong here.'"

———

When I started working on this book, thinking about the mission-driven companies I knew of that had fallen victim to the venture approach, one that immediately came to mind was Good Eggs. Launched in 2011, at the height of both the farm-to-table, slow food trend and the sharing economy, Good Eggs' mission was to challenge the conventional food system by making it as easy to access locally sourced, responsibly produced food as it is to go to the grocery store. Good Eggs would partner with small-scale farmers and artisan producers to connect them with customers in their region who wanted higher-quality food than what they could typically find in a supermarket. Instead of creating a typical marketplace platform, like Uber or Etsy, Good Eggs acted as the intermediary, creating what was essentially an online grocery store. They sourced, warehoused, and delivered all the orders, managing both customer relations with the grocery shoppers and the partnerships with the food producers.

As Rob Spiro, Good Eggs founder and former CEO, explained it to me, there was outsized demand for sustainable food and there was a lot of potential supply in the form of local farms and artisan food makers, but the analog systems that managed this market were too inefficient to scale to the point where they could have a real impact on how we produce and eat food. Spiro hypothesized that if they could use technology to solve those inefficiencies, they could grow the sustainable food market from 1 percent of the entire food system to 10 percent, making meaningful positive change for the planet, our health, and local economies.

Spiro, who sat somewhere on the spectrum between Aligned Capitalist and Conflicted Altruist when he started Good Eggs, had already sold one software startup to Google. He was excited by the opportunity to apply the methods of Silicon Valley to upend the food system, creating value for both the environment and public health in the process. To enable the seamless experience that customers were expecting, Good Eggs had to invest significant resources into building the software-enabled system that could manage the complicated logistics the endeavor required. That meant Spiro and his cofounders would need to raise venture capital if they hoped to achieve their lofty vision.

"I wanted local food systems to grow massively because I really believed it was better for farmers and eaters and local communities," Spiro told me. "And to start a grocery store in San Francisco . . . didn't seem ambitious enough to me. If we innovate and create a new model that works, that uses technology to change the market dynamics, shouldn't that be applied everywhere?"

Spiro's passion to solve this logistics problem with software was infectious. That, along with his track record as a successful startup founder, made it easy to get venture capitalists on board. He was a credible pitchman for something he truly thought could be a massively scalable business that would change the way people ate around the world. "We were the belles of the ball," he told me, about his ability to raise venture capital. In its early rounds, Good Eggs raised money from some of the top firms in Silicon Valley.

While the company grew quickly in its early days by an impressive (even by VC standards) ten times per year, it soon hit a wall. Spiro and his cofounders began to realize that applying the same approach they took to scaling their first business, a software company, in the analog world wasn't going to work. "What we didn't realize is all of the implications of building a business with variable costs of 90%, vs. variable costs of, say, 30%," Spiro said. Unlike in tech, "in the grocery industry (and logistics industries), the marginal costs approach 90%. And what we didn't realize was all of the subtle ways that this difference changes the way a business needs to be run, the way innovation can be managed, the impact of rapid iterations and change."

As Spiro points out, the economics of software are such that the cost of reaching scale is relatively low because once the software is built, it requires very little maintenance and can be distributed basically for free over the internet. The methodologies used by current venture capitalists were developed with software companies—and the ease of distributing software—in mind. This made venture investing even more lucrative because companies could reach massive scale, and investors could reap huge rewards, for a relatively small amount of up-front capital compared to the value these startups were capable of creating.

But as venture capital entered its world-eating phase in the late 2000s, it began applying that methodology to businesses that weren't fundamentally software companies. When you think about the worst examples of Silicon Valley malfeasance—companies like WeWork, Theranos, and Juul—they are all examples of where the methods that worked for venture-backed software companies were applied to physical-world products or services. Once it became clear to the founders and the investors that these businesses wouldn't scale like a software business, they started cutting corners in an attempt to defy that reality. Chaos ensued, and most of the time the rest of us were left to pick up the pieces.

A company like Good Eggs that operates in the physical world, needs stuff—delivery trucks, warehouses, refrigerators, packing materials—for every dollar of revenue it earns. All that stuff amounts to variable costs, which eat into profit margins. Soon enough, the tensions between the venture methodology Spiro and his investors were applying and the realities of the kind of business they were actually running began to emerge.

Shortly after they raised their first real round of funding, Good Eggs did what you would expect a software company—but probably not a company that sells groceries—to do: they expanded their services very quickly to markets across the country. "We were running a business that was quite unprofitable, but improving, with the goal that if we continued at this rate of change, and reached a certain scale, we could reach breakeven." But soon Spiro, his cofounders, and the investors came to understand that there were flaws in the model. The skyrocketing growth they had experienced in the beginning stopped. They had a choice: expand to more cities so they could raise more venture and hope that they could buy enough time to make the economics work; or retrench to San Francisco. The second option would allow them to extend their cash on hand for much longer, but it probably meant they would have to give up any hope of raising the capital it would take to fulfill their world-changing mission. So, Spiro decided to stick with the expansion plan—a plan that also allowed him to keep innovating on the business model—and use that as a selling point as he went to pitch more funders. By late 2012, barely a year after the company launched, they were pushing forward with plans to operate in Los Angeles, New Orleans, and New York alongside the growing Bay Area business.

The cracks widened. In 2013, shortly after the company opened for business in Los Angeles, they were forced to shut down a warehouse for a couple of months to fix a refrigeration issue flagged by the local health department. When Spiro raised the issue at a board meeting, his investors told him to lay off the workers while the warehouse was closed. Spiro was taken aback. He felt he owed it to the workers to keep paying them while they fixed the problem. "It'll cost us some money," Spiro says he told his investors in that meeting. "But when we reopen we have people on the team who are loyal and motivated. And also it's the right thing to do." His investors disagreed. Ultimately, Spiro pressed the issue (making a decision to push back against his investors in a way that a founder with less experience and cachet may not have felt confident to make) and won that battle, but he lost a lot of goodwill from his board. He thinks it planted a seed that eventually led to his leaving the company two years later.

The divergence between the software-driven approach his investors wanted to take and the logistics business that Good Eggs was becoming was wearing on Spiro. "The part of the software approach where you iterate quickly, and if it doesn't work, you change it. Continuous improvement. All these things are nice ideas," he said. And they come at a low cost. Rewriting software to tweak a product is relatively simple and usually doesn't require a massive shift in operational capacity. On the other hand, "When you have a logistics business [like Good Eggs] and you need to change things, that means you need to fire people, close a warehouse, retrain people, change their jobs." These workers were mostly lower-wage service workers and manual laborers who didn't have the cachet and in-demand skills that most tech workers have. When software jobs are eliminated, tech workers tend to be snatched up by other hungry and growing tech companies. But when warehouse and delivery jobs go away, the human toll is much higher. That weighed heavily on Spiro. "It totally wore me down. Having to fire people was super traumatic, for me and for them. I'm talking about firing people for no fault of their own because we had to change the model and adapt." His investors paid lip service to what he was feeling. "They'd say 'we understand this is hard for you Rob, we're here for you.' But if I proposed 'well maybe we don't have to do this, maybe there's another way even if it's less profitable' they said no."

As Spiro was preparing to leave Good Eggs in 2015, he had accepted that the company would not be the world-altering success he hoped it would be. "If you can continue to tell a story that you're on a massive growth trajectory, you can probably find investors who are willing to believe that story. I had investors on my board who realized I wasn't on that trajectory." At that point, his investors wanted the company to pivot from the growth-at-all-costs approach to turning a profit so that they could make Good Eggs look more appealing as an acquisition target. The ideal end goal in their minds was to get acquired by a large supermarket chain, which they couldn't do if they were losing $10 million a year and only growing by 30 percent. So they did the thing that Spiro never wanted to do. They shuttered the expansions and scaled the company back to the San Francisco Bay Area, leaving hundreds of employees out of a job and their other partners, such as local food producers, in the lurch.

Looking back now, Spiro realizes that his approach—to bring a software mindset to a food system business—was misguided. He learned a lesson that many entrepreneurs I spoke with also learned the hard way: not every problem needs a venture-scale, software-driven solution. "You have to pick one or the other," he said. "If you want to solve a specific problem you have to use whatever tool is necessary to solve that problem. And if you want to use a specific tool [like software] then you've got to find a problem that that's the right tool for. I tried to hold both of those points fixed at the same time. Maybe I should have just held on to the tool and found a different problem or maybe I should have held on to the problem and just built a grocery store."

That's ultimately what Good Eggs has become: a really great grocery business. As of this writing, still operates in the Bay Area; I myself am a satisfied customer. I think it's a pretty happy ending as far as these things go, although investors would probably disagree with me. After its retrenchment in 2015, the company raised $150 million more in two separate funding rounds and expanded back to Southern California in 2021. By early 2023, as the Covid-driven grocery delivery bump had faded and interest rates were on the rise, Good Eggs' valuation was down 94 percent from where it was at the end of 2020. Most investors are on track to lose most of what they put into the company. It is fair to say Good Eggs is no longer the belle of the venture capital ball.

And yet, the product and the service seem to be working great. And the Good Eggs warehouse and delivery workforce has much higher job quality standards than the average food service company, including generous health benefits, paid leave, and equity in the company. Eight years later, Spiro is reflective: "What we succeeded in doing is using more than $100 million in venture capital to build a great, idealistic grocery store for the Bay Area. [Good Eggs has since reopened in Los Angeles.] The investors might lose all their money but that $100 million went into the local food ecosystem in the Bay Area. That's cool. That's a nice way to see the story."

LocalData and Good Eggs, while they had two very different experiences with venture capital, both demonstrate the cost to entrepreneurship when startups contort themselves to fit into the traditional VC model. Companies that aren't meant for venture-scale growth, either because their market is too small or the industry they work in doesn't fit the VC mold, become distorted. This distortion has a variety of consequences: missed opportunities to address real problems; burned-out and disillusioned entrepreneurs whose energy and talent are sacrificed in the push for ever-greater returns; employees and customers who get the rug pulled out from under them. Unlike the stories of venture capital and tech excess that make headlines, these stories—which are far more common than the big frauds—represent the hidden cost of venture capital's world-eating mentality.

It isn't just the individual companies that become distorted. Venture capital also distorts whole sectors by determining which types of companies survive and which don't. By pumping enormous amounts of money into companies that promise to grab as much market share as quickly as possible, venture capitalists essentially choose winners. In individual markets, where there may be a variety of companies employing different business models—some that may be more sustainable and offer better societal outcomes—the venture-backed model will be the one to succeed simply because it has access to more resources than the others. The lack of capital options for different kinds of startups ensures that the venture-style model will almost always prevail.

One of the sectors where this dynamic can be seen most acutely is the emergent home cooking industry. Home cooking businesses—which have long existed as part of the informal economy, usually run by immigrants or people of color in marginalized communities—began to gain mainstream attention during the sharing economy's heyday. Platforms emerged that connected home cooks with diners who ordered their food in the same way they would from DoorDash or Postmates.

One pioneer of this sector was Matt Jorgensen, who cofounded a company named Josephine in 2014. Jorgensen, a Conflicted Altruist, was motivated to help people who had been left behind by the formal economy to start their own businesses. He and his cofounders came to the conclusion early on in their partnership that using technology to create a marketplace where home cooks could find their customers was going to be the best way to accomplish this goal at a meaningful level.

One of the defining characteristics of software-based marketplace businesses is that they become more attractive as more people join the platform, a phenomenon known as network effects. As Josephine tried to attract customers, it would have to ensure there were enough home cooks on the platform for customers to choose from. And conversely, to attract home cooks, it would need to show that there were enough customers to make it worth the cooks' time. This kind of chicken-or-egg problem can be difficult to solve, but once a company cracks the code, it produces what venture capitalists like to call a flywheel effect: customers and providers joining the platform in a self-reinforcing upward spiral that drives enormous scale. As Jorgensen and his cofounders were starting Josephine, Uber was busy proving that the best way to achieve this flywheel effect for a marketplace business was to throw lots of money at it, so Josephine figured they had no choice but to raise venture capital and deepen their pockets.

They had their reservations about this approach. "We thought about, is this a nonprofit? Or maybe a cooperative?" Jorgensen told me. "But to create the multi-sided network we felt like we needed to make this successful, the risk-tolerant capital was just not there in the traditional philanthropic or cooperative worlds." They grudgingly started fundraising and naively hoped they could find the best of both worlds. They wanted investors who had high tolerance for risk and could move money quickly but who were going to let them run the company the way they wanted, leaving them in full control of how the business was shaped. "We thought we were cleverer than the system. Our goal was to get [Josephine] to a certain liftoff stage without having sold a majority of the company, and then never have to raise [venture capital] again," Jorgensen said ruefully.

What actually happened to Josephine was much different. They launched in the Bay Area in 2014, amid the gauzy optimism of the onset of the sharing economy era. Setting up a hot-food business that was run from a home kitchen was explicitly illegal everywhere in the US, but then again, so was operating a taxi service with privately owned vehicles. Unlike Uber, however, Jorgensen thought that by developing a relationship with regulators, he could convince them to look the other way. This, as it turns out, was a miscalculation. His efforts only succeeded in bringing the company to the attention of local health departments before they had gained enough traction to sustain a regulatory battle. Shortly after Josephine started operating, they were forced to shutter their operations in California altogether.

Pulling from the lessons they learned in California, they decided to take a more strategic approach as they entered their next two markets: Seattle and Portland. They enlisted the help of Airbnb's first head of policy, who advised them to launch right away and, in parallel, unleash a charm offensive on key decision-makers. They turned the mayor of Portland into an evangelist for their cause and con-vinced him to lobby state lawmakers to legalize home cooking. In the state of Washington, treating home cooking's legality as an open question rather than a settled fact, they advocated for the creation of a task force (which was ultimately established) that could make recommendations on how to regulate the practice. At the same time, they were building a grassroots movement of home cooks and other allies who could pressure lawmakers from the outside not to crack down on their growing businesses.

Despite these efforts, they didn't have much more success than they'd had in California. The task force in Washington ultimately decided not to recommend legalizing home cooking in the state. And, to add insult to injury, one of the task force members went so far as to file a complaint against Josephine with the Attorney General's of-fice. At risk of exposing the company, and all the cooks who were using the platform, to legal liability, Josephine's founders decided they needed to pull out of their Seattle market as well.

During this period of tumult in the Pacific Northwest, Josephine began organizing a coalition of home cooks and aligned advocacy organizations back in California to lobby to make home cooking le-gal there. Jorgensen began spending more and more of his time on this effort, seeing it not just as critical to the long-term success of the business but also the right thing to do for the many cooks who had stuck with them through the hard times. "We had a lot of conversa-tions about how to continue supporting the cooks, because we had made a lot of promises to them that this wasn't just a startup, this was a movement," Jorgensen said. "It was clear at that point if we just walked away it would be letting down the hopes of something that was much larger than Josephine."

Their investors weren't thrilled, to say the least. "They were like 'why don't you just put your heads down, it seems like you're putting a lot into this political stuff and these guys over here . . .' [indicating Uber with a wink and a nod]," Jorgensen paraphrased. The Josephine founders decided to ignore their investors and forge ahead with the campaign anyway, hoping those investors would come to see that the best path forward was to legalize home cooking before they created any more legal risk for the company or its cooks. It ended up being a fatal choice for Josephine. Their investors remained convinced that the Uber approach was the only way to deal with regulators, and Josephine wasn't able to raise the money they needed to get to the other side of the regulatory fight. It shut down in 2018.

It wasn't all bad news, though. That fall, their organizing efforts paid off. They passed the Homemade Food Operations Act in the California legislature, and Governor Jerry Brown signed it into law a few weeks later. They did so with the help of another home cooking entrepreneur, named Akshay Prabhu, who started a company called Foodnome around the same time.

Prabhu is a neuroscientist by training who, since childhood, has fostered a dream of owning a restaurant. In 2017, when he was working at a hospital in Sacramento, he decided he was going to pursue this dream on the side. He opened a café in his garage that became popular enough to gain some local media attention, which was great for attracting customers but also put him on the radar of the local health department, which made him shut it down. That experience motivated him to find a way to make it easier for people to start home-based food businesses. That's when he connected with the Josephine founders and got heavily involved in the movement to legalize home cooking in California. The movement the three of them helped start evolved into a nonprofit, the COOK Alliance, that supports home cooks in setting up legal businesses run out of their kitchens. It also advocates for the adoption of similar laws in states across the country. As a result of the COOK Alliance's advocacy efforts, in 2021, Utah became the second state to legalize home cooking businesses. Jorgensen remains a board member.

After the Homemade Food Operations Act passed in California, Prabhu moved to Riverside County in the Inland Empire. Riverside was the first county to adopt the provisions of the new law, which required counties to opt in in order for home cooking to be legal, and Prabhu saw a business opportunity to help home cooks get up and running. He knew he would need funding to get started and, seeing no other options, set out to raise venture capital. It was important to him that Foodnome operate completely aboveboard, prioritizing a commitment to food safety while supporting the growth of their cooks' business aspirations, and he tried to make that a competitive advantage when he pitched VCs. He thought there might be a space for a company to take the opposite regulatory approach from what Uber and Airbnb were doing. He was able to convince a couple of early investors that this might be a compelling opportunity, bringing in $2.5 million to fund his growth. But the number of investors who were willing to take a bet on the legal approach to home cooking was vanishingly small. Foodnome's commitment to playing by the rules slowed them down, making it impossible to meet the growth expectations of most venture capitalists.

Having held his own dreams of owning a restaurant one day, he also placed a premium on giving home cooks the tools to grow their businesses, whether that happened on the Foodnome platform or not. But running the business that way, creating what he called a "business in a box" approach rather than building the marketplace that Josephine had set out to create, limited the growth potential of the company. Trying to convince investors that another model was possible was a losing battle, and Prabhu found it very difficult to raise additional funding to expand the business.

The work that the COOK Alliance had done to legalize home cooking in California attracted other entrepreneurs eager to capitalize on the changing regulatory environment. "As soon as the [Homemade Food Operations Act] passed it got onto a lot of savvy entrepreneurs' radars," Jorgensen told me. Perhaps the savviest of them were Alvin Salehi and Joey Grassia, who launched a company called Shef within twenty-four hours of the act becoming law. Grassia and Salehi had a more typical profile for venture-backed startup founders than the founders of Josephine and Foodnome did. Grassia was a former Facebook employee who had started two food businesses on the side. Salehi cofounded code.gov, a project of the Obama White House that shares open-source government technology projects across the US government, and is a venture investor himself.

Grassia and Salehi are both the sons of immigrants, and they lean heavily into that fact when explaining how they came to found Shef. According to them, their mothers were always cooking at home, infusing in them a spirit of togetherness and family connected to home cooking. They have placed the immigrant experience at the center of Shef's marketing pitch, proclaiming:

> Shefs are aunties and abuelas, immigrants and refugees, stay-at-home parents and restaurant dreamers. Together, they represent countries around the world, from Algeria to Korea, to India and Venezuela. And the best part? Shefs are your neighbors, passionate local cooks in your community who are doing what they love most—sharing their food and culture with you.

Shef claims that as of mid-2023, 85 percent of the cooks on its platform are women, 80 percent are people of color, and they represent eighty-five countries.

Both Foodnome and Shef applied to be a part of Y Combinator's Winter 2019 class. Only Shef was selected, perhaps because Shef was able to tell the story of venture-scale success that Foodnome could not.

Neither Salehi nor Grassia would talk to me on the record, so I don't know what, if any, tensions they experienced when they began raising the money they needed to scale. Were they like Rob Spiro, who saw the venture approach to growth as fully compatible with his vision for Good Eggs? Or were they more like Matt Jorgensen, who had misgivings about venture capital but ultimately was most interested in building a scalable business so was willing to go along? It's impossible to guess.

What is clear is that the decisions they have made as they constructed Shef often stand in stark contrast to the decisions that Josephine and Foodnome made. Their decisions have gone down very well with investors. By the middle of 2023, Shef had raised five rounds of funding at a total of over $100 million. Shef is among the hottest of startups, boasting investors like Andreessen Horowitz, who led its Series A round, to celebrities like Katy Perry and Odell Beckham Jr. In 2021, at barely three years old, Shef was added by *Forbes* to their list of next billion-dollar startups.

Salehi and Grassia have managed to achieve this astronomical success in a regulatory context that hasn't changed much since Josephine went out of business. The Homemade Food Operations Act in California requires counties to opt in, and as of this writing, only fifteen have done so. In the rest of the state, and every other state except Utah, home cooking businesses remain illegal. But that hasn't stopped Shef from expanding across the country. How did they accomplish all this when others couldn't? First, they seem to be willing to take the Uber approach to entering new markets. In 2023 they were already operating in eleven states and Washington, DC, when they announced plans to expand nationwide after their latest round of funding. Shef is ostensibly justifying its expansion in states where home cooking isn't legal in one of two ways. The first, they might argue, is that cottage food laws, which allow home food businesses to sell nonperishables like jams and granola, apply to their cooks. The second is that most states allow "home cooking" as long as the food is prepared in a commercial kitchen. Shef's website is quite vague about what this means for home cooks, saying only that the means of getting food to customers "varies in each market based on local requirements." Shef partners with third-party delivery services like DoorDash which, we must assume, pick up the food at the commercial kitchens. But whether the meals have actually been prepared in the kitchen, rather than cooked and assembled at the home cook's home and then brought to the commercial kitchen for pickup, is unclear.

Like Jorgensen, Grassia and Salehi knew that, to build a useful product that would keep customers coming back, they needed to achieve network effects. But Jorgensen and his cofounders wanted Josephine to be more than a transactional platform that felt much the same as impersonal delivery apps like DoorDash or Postmates. They wanted Josephine to facilitate connections between neighbors who lived in the same communities, breaking down divides and strengthening the social fabric. This commitment to community building meant that if they wanted to connect neighbor with neighbor, they would essentially need to build many neighborhood-sized markets instead of a single metropolitan-sized one, upping the degree of difficulty significantly on achieving network effects. They also eschewed delivery, preferring that diners pick up their meals to facilitate a face-to-face interaction, a decision that seriously irritated their investors and became a major sticking point as they tried to raise more money. It also complicated their ability to operate legally in most any market.

The distance between the cooks and the customers isn't just an unfortunate side effect of Shef's effort to toe the regulatory line. It also serves to keep the cooks reliant on Shef—a critical element of achieving venture-scale growth. To achieve the flywheel, Shef doesn't only have to get users onto the platform very quickly, it also has to keep them there. Shef has made several design choices that keep the cooks reliant on the platform in order to run their businesses, making it harder for them to leave and build their own, independent customer base. For example, Shef does not provide the cooks access to any customer data, meaning if the cook leaves the platform, they also leave behind any information about loyal customers that could help them set up a brick-and-mortar restaurant (something many of the cooks on these platforms aspire to). They also aren't able to export any of their performance data, which would be extremely useful if they wanted to, for example, try and secure a small business loan.

Foodnome takes a different approach. In addition to working with cooks to design their menu and hone their marketing skills, Foodnome helps its home cooks receive all the certifications and licenses it needs to operate their business legally. Foodnome's commitment to operating legally already places a huge cap on how much they can grow, given that the model has been adopted in only fifteen of California's fifty-eight counties. But the certification process is also time-consuming and expensive, which further limits Food-nome's ability to add cooks quickly.

Shef, on the other hand, lays the legal responsibility squarely at the feet of the cooks, creating a risk that the cooks using the platform (many of whom do not speak English as a first language) may not understand they are assuming. Shef's terms of service state:

> Each Seller is solely responsible for complying with all applicable laws, rules and regulations and standards, including but not limited to those pertaining to the preparation, sale, marketing, packaging, handling, and delivery of all Meals ordered through Shef . . . Each Seller is solely liable for the quality, safety, and freshness of its products, and Shef does not verify the credentials, representations, products, services, or prices offered by any Sellers, and does not guarantee the quality of the product or services, or that Sellers or Meals comply with applicable laws.

Shef's marketing materials state that all home cooks are required to pass a food certification exam before they begin selling on the platform, but that exam is not administered by a health authority (after all, the cooks aren't supposed to be using their home kitchens in the first place). And while Shef makes claims about what home cooks are required to do (using a meat thermometer to ensure safe cooking temperature, for example), no one, as far as I can tell, is verifying that the cooks on Shef's platform are meeting these requirements. If a food safety issue does occur, Shef's terms of service make it clear that the cooks alone carry the burden of liability.

This isn't the only risk of exploitation that cooks on the Shef platform experience. Pitched on the gig economy ideal that they are starting a small business, Shef's cooks wind up saddled with all the responsibilities and costs associated with being a business owner, with very little of the support. One Shef cook posted about his experience on Reddit:

> I am fairly new to the venture and feel like I am not making any money. In fact, I am actually losing money. Between the ingredients, supplies, gas (stove), water-bottle ice packs, transportation and especially MY TIME, I am in default. So far I've been taking one for the "biz" in order to establish a ratings base. I have also considered raising my prices, but I am trying to stay competitive since I'm new. I don't want to wander too far from the prices offered by fellow "shefs" or local restaurants either.

Another cook responded to that post:

> I don't think Shef.com necessarily sets you up to succeed. They want you to make a menu and start accepting orders right off the bat. I feel like I was pretty ingredient utilization/price conscious from the beginning and I STILL had a rough first few weeks (days, I only took orders 1 day a week). They're helpful in a lot of ways, but they also gently push you to price your items low, send out free samples, and make a wide and varied menu, which can create a lot of losses for someone, especially starting out.

Shef's low-touch approach allows it to scale much faster than its competitors. "Shef can onboard 100 cooks in the time it takes us to do five," Prabhu told me. That faster growth clip is obviously much more appealing to venture capitalists and makes it much harder for Foodnome to raise the kind of capital they would need to attempt to show that their model can deliver big returns as well. "When they can say they're growing 75% month over month and we can grow 30% month over month, it changes the narrative and makes it harder for us to raise capital."

This dynamic, where venture capitalists reward the startups with the juiciest numbers without asking too many questions or caring about how they got there—or caring what the answers are—happens across the entrepreneurial landscape. In one recent example, Charlie Javice, the founder of Frank, a venture-backed startup that made college financial aid easier to access, was charged with fraud for inventing customer data. She fabricated millions of users, making it look as though her company was much larger than it was in order to tell the story of venture scale that would justify the $20 million she raised from investors. Her fraud was ultimately discovered after Frank was acquired by JP Morgan Chase for $175 million, but JP Morgan Chase wasn't the only victim here. The willingness of venture capitalists to take what Javice said at face value and pour millions into her company created a standard that Frank's competitors couldn't live up to. One of those competitors told *The New York Times*, "When any other person had an idea for trying to solve this [college financial aid] problem and went to a venture capitalist, that venture capitalist would say: 'You're not having that much success. Look at what Frank has done.'"

Shef isn't perpetrating fraud. But they do seem willing to operate in a way that pushes the boundaries of legality, and certainly exposes their cooks to a significant amount of risk. Their willingness to do what's necessary to achieve the kind of venture scale that appeals to investors makes companies like Foodnome look like bad businesses by comparison. The result is that models like Foodnome's don't get the opportunity to prove that a slower, more responsible approach could result in a strong business too. Given the choices Foodnome has made, it's unlikely it could ever be a venture-scale business like Shef. But it could achieve some level of success and deliver returns—not to mention create real economic development opportunities for would-be restaurant owners—if there were investors who saw the opportunity that companies like Foodnome represent.

Ultimately, the lack of investors willing to support Foodnome's model has meant that the company couldn't continue to operate independently. It was folded into the COOK Alliance, which will continue to support Foodnome's cooks as part of its nonprofit programming, in early 2024.

Because I couldn't talk to anyone involved with Shef, it's hard for me to know whether the way it appears from the outside is a true reflection of the conscious choices the company is making. There may be a much more charitable version of the story than the one I'm able to tell. It could be that the Shef team has figured out how to thread a very small needle, aligning the opportunity for scale with what's in the best interests of the cooks. If they have, it would require an almost perfect coming together of the moral wherewithal and savviness on the part of the founders, a set of investors who are particularly patient, and a market that presents few challenges. If this is all true, it would be the exception that proves the rule. This narrow path is possible, but vanishingly hard to stake out in the venture-backed startup world, especially in areas like the one Shef plays in, which is hugely ripe for exploitation.

The entrepreneurs I feature in this chapter have all taken a step back from mainstream entrepreneurship. Alicia Rouault has gone back to her government roots, working in the United States Digital Service at the White House to make it easier for families to sign up for economic benefits. Matt Jorgensen has worked on a variety of efforts, some of which we'll talk about later in the book, to support mission-focused businesses. Rob Spiro has emigrated to France, where he runs an incubator that supports what many in Silicon Valley would think of as lifestyle businesses. In their way, they are contributing to healthy economic growth and the development of a new conception of what innovation can look like in a capitalist economy. But it is telling that to do so, they had to move to the edges of the culture, unable to find a solid foothold for their way of thinking in the mainstream innovation economy. What would it take to make entrepreneurs like them the norm rather than the exception?

3

VENTURE CAPITAL AND LABOR

HOW THE GIG ECONOMY HAS UPENDED THE JOB MARKET

Across the economy, venture-backed business models have been upending decades of critical labor protections provisioned through the New Deal that, alongside access to home-ownership, form the basis of middle-class stability.

While it is best known as the framework that ushered in the American social safety net as we know it, the New Deal of the 1930s also created the standards that would govern the worker-employer relationship. The New Deal created a bargain between employers and workers that remains essentially intact today: workers would offer their labor to power the growth of industry; in exchange (in addition to compensation), employers would assume liability for the worker during working hours and abide by a set of laws that provided workers with basic protections against abuse and exploitation. The safety net benefits provisioned by New Deal policies also hinged on employment status. Social security benefits, unemployment insurance, and, of course, access to health care all tie their access to work history. Those workers who fell outside the legal definition of an employee, those who were deemed to be independent contractors essentially running their own small businesses, were left to provide these services for themselves.

The definition of who counted as an employee and who counted as an independent contractor may have seemed more obvious in the 1930s, but as the economy evolved, these classifications became harder to distinguish. By the twenty-first century, this lack of distinction would come to generate some of the biggest political battles of the digital age.

The case that got the ball rolling, however, didn't have anything to do with technology. Just before Christmas in 2004, Pedro Chevez and his coworkers at a California parcel delivery company named Dynamex came to work to find that the company had unilaterally converted them from full-time employees to independent contractors. This meant, despite Dynamex continuing to determine where, when, and how the drivers performed their roles, the drivers—not their employer—would now carry the burden for all their business expenses. They had to use their own vehicles for deliveries; they wouldn't be reimbursed for mileage; they would no longer receive overtime; and they were on their own for benefits like health insurance and retirement.

The following spring, a full five years before anyone had uttered the term "gig economy," Chevez joined with his coworker Charles Lee to file a lawsuit against Dynamex alleging the reclassification was illegal. Thirteen years later, after a protracted and labyrinthine journey through the courts, the California Supreme Court ruled in their favor, finding that Dynamex had indeed misclassified the workers, opening the door for the workers to file claims against Dynamex with the California Department of Industrial Relations. The fact that Chevez and Lee filed the lawsuit, stuck with it for over a decade, and ultimately prevailed is as close to a miracle as you can get in labor relations. Fighting back against misclassification is an uphill battle for workers, and Dynamex's decision to convert their employees to independent contractors was made knowing they would probably get away with it.

The ability of many freelancers to operate as independent entities is a necessary part of the American economy. But because the rules around who qualifies as an employee and who qualifies as an independent contractor have never been made clear, many employers (like Dynamex) have taken advantage of the murkiness to convince their employees that they are actually independent contractors. This type of misclassification has proliferated across the economy. According to the Department of Labor, up to 30 percent of US employers have misclassified employees as independent contractors. But the Dynamex decision took a significant step toward protecting workers against misclassification. In addition to ruling in favor of Chevez and Lee, the court also outlined a new, simpler standard for determining classification status, called the ABC test. According to the test, workers could be classified as independent contractors only if they (a) were in full control of where, when, and how they performed the work; (b) were performing work that was outside the core business of the client for whom they were working; and (c) were established as a business entity (i.e., they were incorporated or had multiple clients).

Most workers don't know there is a distinction between an employee and an independent contractor, or that there is any way to challenge their employer for misclassifying them. If one does decide to bring a case, they usually get mired in a drawn-out legal battle while lawyers fight over technicalities and precedents. In the end, the majority of workers who are misclassified never receive justice.

When the Dynamex decision came down, labor unions were jubilant. "When the ruling dropped it was like a bomb dropped in this war we'd been waging for 40+ years," Doug Bloch, a political operative working for the Teamsters in Northern California at the time, told me.

But for many in Silicon Valley, the ruling was ominous. Despite the court's decision being based on facts from a case that was first filed well before the rise of the gig economy, it was widely understood by those in tech and the labor movement that this ruling would have profound consequences for companies like Lyft and Uber. "From the moment Dynamex passed, people were saying 'oh, you know who's going to see this as a threat,'" Derecka Mehrens, who was at the time executive director of Working Partnerships USA, told me. Gig companies had been deflecting criticism from labor advocates for years. But the Dynamex case—and the ABC test—was the first real threat to their business model, which hinges on their ability to classify drivers as contractors.

State Assemblymember Lorena Gonzalez, who got elected to the California state legislature after a long career as a union organizer in the San Diego area, saw an opportunity. While the court decision created a legal precedent, she wanted to strengthen its meaning by codifying the ABC test into law. In 2019, she authored Assembly Bill 5, which would do just that.

Almost immediately, AB5 became one of the most contentious

legislative battles in California history. Freelancing is deeply embedded across the economy, and many industries from trucking to entertainment have been designed to accommodate that way of working. To many freelancers, AB5 risked upsetting a system that was working for them. Ultimately, many professions managed to get themselves exempted from the rules. But the most formidable opposition came from the venture-backed gig platform companies, who were squarely in Gonzalez's sights, particularly those who relied heavily on drivers and delivery people to make their businesses work. Uber, Lyft, DoorDash, Postmates, Caviar, and Instacart formed a coalition to ensure that the rules laid out by AB5 never applied to them.

The gig companies argued that forcing them to classify drivers as employees would harm both workers, who would lose flexibility, and customers, who may experience longer waits and more expensive rides. But according to many labor experts, this argument was a smoke screen that obscured the true reason they perceived AB5 as such a threat. The tenuous gig business model depends on being able to shift the costs of doing business—insurance, vehicle costs, workers' comp, health care, and paid medical leave among other things—to the people doing the work.

With the future of the bill in the balance and recognizing that this showdown between labor and business would have huge political ramifications for years to come, representatives from labor and the gig companies began a series of closed-door meetings to try to find a compromise. There was some common ground (some representatives from labor, for example, were open to the concept of a third classification status other than employee or independent contractor, as long as it allowed workers to organize), but ultimately the negotiations failed. The major labor unions went on to support AB5 and the bill was signed into law by Governor Gavin Newsom in September 2019.

Obviously, the gig companies saw AB5 as an existential threat. But why? The conventional wisdom among labor advocates is that the gig companies don't want to carry the additional cost associated with employing the workers directly. That's partly right, but the threat to the gig economy's financial model goes much deeper than that. Employers in every sector have been trying to wriggle out of their end of the labor bargain for decades. What's unique about the gig industry's reaction to AB5 is that the bill struck right at the heart of their identity.

There are a handful of business models that are well suited to achieve venture-scale returns. The marketplace business model, like the one employed by Shef, is one of them. In a marketplace—or platform—model, the company makes money simply by connecting buyers and sellers. It doesn't hold its own inventory. It doesn't manage any of its own logistics, like getting products to customers. It simply connects supply and demand, taking a cut of the transaction when it happens. The simple elegance of the marketplace model, and the fact that it requires very little overhead, makes it perfect for achieving massive scale in a short period.

The ABC test encoded in AB5 threatened that marketplace model, at least as it applied to gig delivery and transportation companies that relied heavily on drivers. One of the components of the test is that workers can be classified as independent contractors only if they "perform work that is outside the usual course of the hiring entity's business." According to the gig companies, as marketplace businesses, the usual course of their business is building software, not providing transportation (or other courier services). When AB5 passed, Uber's chief legal counsel, Tony West, told CNBC, "Several previous [court] rulings have found that drivers' work is outside the usual course of Uber's business, which is serving as a technology platform for several different types of digital marketplaces." According to this logic, the workers who comprise the core of Uber's business are the software engineers who keep the platform up and running, not the drivers who ferry riders from one place to another.

Since at least 2012, Uber and the other venture-backed rideshare companies have been proclaiming publicly that they are not transportation companies but rather software companies that run a marketplace where people who need rides connect with drivers who can take them somewhere. Uber's terms of service at the time of this writing state **"YOUR ABILITY TO OBTAIN TRANSPORTATION, LOGISTICS AND/OR DELIVERY SERVICES FROM THIRD PARTY PROVIDERS THROUGH THE USE OF THE UBER MARKETPLACE PLATFORM AND SERVICES DOES NOT ESTABLISH UBER AS A PROVIDER OF TRANSPORTATION, LOGISTICS OR DELIVERY SERVICES OR AS A TRANSPORTATION OR PROPERTY CARRIER."** (Their caps and bolding.) It's clear how seriously Uber and the other gig companies take this: in their view, the drivers are small business owners who use Uber's software tools to find their customers. Companies like Uber, Lyft, and Postmates merely facilitate the transaction.

While the gig companies claim that drivers have full control over their working environment, the reality is very different. The gig companies are responsible for setting the terms of each driver's engagement. For example, small business owners are able to set their own prices, but in the gig economy, the fare each rider will pay is set by the companies. The platforms also employ systems that manage the work of the drivers and can fire them if they don't meet expectations. Uber has famously banned drivers from its platform whose rating dips too low. Each of these powers over the working conditions of drivers would seem to define drivers as employees of these companies.

Marketplace business models are lucrative because they are simple. By operating in this way, as Hoffman and Yeh put it in *Blitzscaling*, "online marketplaces avoid many of the growth limits of human or infrastructure scalability." In the logic of venture-backed start-ups, employees are overhead ("growth limiters" as they are called in *Blitzscaling*) that not only slow startups down but also threaten their very identity as companies that are worthy of venture-scale investment in the first place. If the drivers are employees of the companies, providing a service on behalf of Uber or Lyft, the marketplace model falls apart. Suddenly, instead of the companies being enablers of a transaction, they become purveyors of a product or service, essentially no different from a taxi company, and taxi companies can't achieve venture-scale growth or command the high valuations that come with it.

It is through this lens that Silicon Valley was watching the AB5 drama play out. A handful of gig companies being forced to reclassify its workforce wouldn't just be a hit to the balance sheets of those individual companies, it would call into question a core tenet of the venture playbook and put at risk the valuations of dozens of other venture-backed businesses that relied on it. It would make it harder for investors who already had stakes in marketplace companies to convince future investors that these businesses were as valuable as they claimed. If Uber was just a taxi company, then Shef might be just a takeout joint and Good Eggs just a grocery store. The idea of Uber becoming just another taxi company isn't a threat to Uber and its investors alone. It is a threat to a cornerstone business model that enables power law outcomes for all venture capitalists.

For these reasons, for gig economy companies who sold themselves to investors as marketplace platforms, AB5 was the grim reaper.

Less than three months after Governor Newsom signed AB5 into law, the gig platform companies used their significant political and financial capital to make good on a threat that had been hanging over the compromise talks: they filed the paperwork for a statewide ballot initiative that proposed a third classification status that would apply only to gig drivers and delivery people. In return for providing a guaranteed minimum wage and establishing a shared fund for benefits like health care and paid sick leave, the companies would be absolved of all legal responsibility for their drivers. The initiative, which was named Proposition 22 on the ballot, outlined that drivers would get paid only when they were actively driving or on the way to a confirmed pickup or delivery. A 2020 study by researchers at UC Berkeley Labor Center found that the actual minimum wage that drivers would receive under Prop 22, when you took into account the time spent waiting for rides and driving between jobs, was $5.64 per hour. Prop 22 also included a clause that required any amendments to the proposition, once it became law, to receive a ⅞ths majority in the legislature, an astonishingly undemocratic power grab on the part of the companies. "[The gig companies] wrote and bought the law they wanted," said Mehrens. "[Labor representatives] advanced AB5 with no exemptions for gig corporations. Proposition 22 then cemented a bad deal for workers, enabling those corporations to maintain an exploitive business model that extracts profits through denying their workers access to the basic labor protections most workers are entitled to under US law."

In November 2020, Prop 22 passed by eighteen points. Uber, Lyft, and the other gig platform companies contributed more than $200 million to the Yes campaign—ten times more than the No campaign spent—making Prop 22 the most expensive ballot measure in American history up to that point. Their deep pockets allowed them to overwhelm the airwaves, stuff mailboxes, and buy the endorsement of interest groups like the California NAACP. They also tapped into the massive reach of their products, embedding campaign messages that appeared whenever customers or drivers opened their apps. Uber was even sued by a group of drivers who felt threatened by the in-app messages they received, which they felt pressured them into voting for Prop 22. Instacart forced its delivery people to hand out Prop 22 campaign materials to customers.

While Prop 22 was a huge victory for the gig platform companies, it was only one battle in a much longer war. California labor groups sued to have the Proposition ruled unconstitutional. After a long court battle, in 2024 the California Supreme Court upheld the ballot measure, allowing gig delivery workers to be classified as contractors permanently.

The companies aren't satisfied with their win in California and are determined to funnel more money into writing the rules that suit their businesses, undermining hard-won labor protections in the process. Prop 22 is quickly becoming the launching pad to establish a new legal framework to govern gig work across the country. Uber, Lyft, and other gig economy companies have launched ballot initiative campaigns in Massachusetts and Illinois that are identical to Prop 22, and they are prepared to spend whatever it takes to get them passed.

Prop 22's impact on the labor force doesn't end with gig workers. Employers in other sectors are looking at the gig industry's political strategy and thinking "How can I make this work for my employees?" After Prop 22 passed, the grocery store chain Albertson's, which also owns Von's and Safeway, eliminated its delivery driver positions and replaced them with independent contractors. In June 2021, Albertson's announced it was partnering with DoorDash, one of the main sponsors of Prop 22, to provide their grocery delivery service from over 2,000 stores across the country. In January 2022, a ballot initiative with unnamed backers was filed with the State of California that would create a Prop 22–style regime for health-care workers who provide services through platforms like BetterHelp. According to Mehrens—and other labor advocates—this is the real threat to labor standards in the US. "The product they've built, fundamentally, is not new," she told me. "Drivers and couriers are not new jobs. What's new is the business model. They are building a set of business, employment and consumer practices—ones that undermine workers—that can be exported to any sector."

Labor experts often point to a strike by air traffic controllers in the early 1980s as a turning point in labor relations in the US. Members of the air traffic controllers' union went on strike in 1981, hoping to win higher salaries and a shorter workweek. The effort backfired. President Reagan broke the union, staffing air traffic control jobs with members of the military. This event had disastrous consequences for labor in the decades to come. Up to that point, labor relations were governed as much by norms as they were by laws. Employers bargained in good faith with their unions in large part because they didn't want to pay the cultural cost of breaking strikes. By busting the air traffic controllers' union (and paying very little reputational cost in the process), President Reagan provided a template for business leaders to follow. Over the subsequent decades, labor unions, and the rules that protect workers, have been eroded to their lowest point since the New Deal came into effect. Prop 22 could create a similar environment, where other sectors are emboldened and decide that with the right narrative, enough money, and the bully pulpit, they can rewrite the rules that govern their workforces too.

Labor activists certainly see the writing on the wall. While Prop 22 applies only to delivery drivers, venture capitalists are expressing hope that its blueprint can be applied across a whole range of jobs. Shawn Carolan, a partner at Menlo Ventures in Silicon Valley (one of Uber's early investors), wrote an op-ed just after the 2020 election hailing Prop 22 as "a way forward for redefining the relationship between tech companies and labor more broadly." He went on to say, "Workers in all sorts of industries—from agriculture to bookkeeping—could benefit from the structure that Prop 22 provides, if it were extended to them."

There are cases where the marketplace business model, so enthusiastically adored by venture capitalists, is a boon to workers. Those that cater to tradespeople or artisans like Thumbtack or Etsy may actually provide more opportunity for sole proprietors whose jobs actually do function more like small businesses. But for platforms that rely on low-wage service work, the risks of exploitation are high. The marketplace platforms, flush with billions of dollars of venture capital, are able to scale the exploitation of those workers in ways that haven't been possible in the past.

The role that workers play in marketplace businesses—as units that enable scale rather than as critical assets of the business—has shaped a perspective within Silicon Valley about the role of labor in the economy that goes far beyond just gig work. While the plight of gig workers tends to grab the most headlines, there is another class of contract workers at tech companies, operating under the radar, who arguably have it worse. Tech companies are driving a massive expansion of "contracting out," hiring temporary and contingent workers through third-party vendors and staffing agencies to fill roles at the company. Unlike gig workers who work *on* the platforms, these contract workers work *for* companies across the tech sector. These workers are legally classified as employees who are the legal responsibility of the third-party agency or vendor and are not independent contractors. This allows tech companies to avoid the misclassification argument altogether.

While the tech industry didn't create the practice of temporary employment, the venture-driven imperative to keep head counts low and labor overhead minimal has made it standard operating procedure across the industry, with serious repercussions for the workers who fill these roles.

Temp work, and the associated rise in staffing agencies, emerged as a mainstream part of the economy in the years following World War Two. As the postwar economy boomed, temporary workers hired through staffing agencies were a useful stopgap to fill the labor needs of growing businesses. Temp work was pitched to women who had "housewifeitis" as a way to get out of the house and round out the household's income. They were called Kelly Girls, after one of the pioneering staffing agencies, Kelly Services. Over the next few decades, as employers came to realize the benefits of having a set of workers who could easily be hired and fired, the practice became mainstream and expanded to include other demographic groups. Aside from small dips during the Great Recession and in the first year of the coronavirus pandemic, the staffing and recruiting sector has grown steadily in recent decades. According to the American Staffing Association, the number of workers in temporary jobs doubled between 1990 and 2008, with an increasing proportion of them working in high-skill jobs. From 2003 to 2019, the size of the staffing industry also doubled.

Staffing agencies present a compelling value proposition for fast-growing companies. They provide an off-the-shelf workforce that has already been vetted and trained, saving startups the time and expense of recruiting and upskilling on their own. Staffing agencies keep those workers on their books as employees so that the client companies don't have to deal with the overhead or logistics around acting as an employer. It's not clear what kind of profit margins staffing agencies make, but TechEquity's research into the practice shows that agencies often benefit from information asymmetries in the market. The client companies who hire through staffing agencies often don't know how much the workers are actually paid. They sign a contract with the agency at a certain rate, and the agency takes an undisclosed cut before it pays the worker. Though some companies have begun to enforce minimum standards for pay and benefits (Microsoft, for example, requires all staffing vendors to provide paid sick leave), these supplier codes are setting a floor well below what the client companies provide to full-time employees. Some of them set standards that they don't live up to. Google has promised that all workers hired through third parties, who Google refers to as "temps, vendors and contractors," or TVCs, will make at least $15 per hour. But data collected by the Alphabet Workers Union in 2023 found that many TVCs were being paid significantly less than $15 per hour. Since that data was uncovered, Google has decided to eliminate their standards rather than enforce them.

The tech industry didn't invent the practice of contracting out, but it did supercharge it in a way that amplified its harms. In 2000, Microsoft settled a lawsuit brought by temp workers who argued that they should have access to the same perks and benefits as full-time employees of the company.

Microsoft initially hired these workers, who the company referred to as permatemps since their appointments were not temporary in practice, as independent contractors, but that raised eyebrows at the IRS. Soon, Microsoft realized they could use staffing agencies as a pass-through entity, but doing so meant the workers weren't technically employed by Microsoft even though they were performing the same roles as full-time Microsoft employees. As a consequence, the permatemps missed out on perks and compensation the full-time employees received, including lucrative stock options that were a core part of the compensation package. Microsoft ended up paying out $97 million to make up for the stock compensation that the permatemps had missed out on.

Despite the settlement, Microsoft didn't convert the workers to full-time employees. Instead, they created even more complicated workarounds that kept them just on the right side of the newly established legal precedent set by the case. "Ironically," the American Staffing Association said after the settlement, "the Microsoft case provides guidance for addressing the issue of when a staffing firm customer may be considered the employer of staffing firm employees."

In other words, now the staffing industry had a playbook to help more client companies offload their workforces onto temp agencies. The practice has since exploded. While there is little data available about the size of this workforce in the tech industry, a few clues have leaked to the press. At the time of the permatemp settlement in 2000, Microsoft had 42,000 full-time employees and 5,000 temps. Twenty years later, *The New York Times* has reported, Google had 121,000 temp workers compared to 102,000 full-time employees, and companies across the industry are following suit.

These new employment practices worked for the hiring companies, allowing them to continue adding more capacity to their operations without adding more head count to their books, but made working conditions even more untenable and precarious for the workers. On paper, these jobs might look like decent jobs. A US-based content moderator, one of the roles that is often farmed out to staffing agencies, could make anywhere from $30 to $75 an hour. But those figures obscure the true cost of these jobs to workers. In practice these positions are extremely precarious and come with other risks that even those who are explicitly engaged as independent contractors don't face.

Chang Fuerte (who uses they/them pronouns) started working as a contractor at Google's Shopping Express division right out of college. They were recruited at an on-campus job fair by Adecco, a multinational staffing agency that contracts with Google and many other tech companies to provide what Adecco refers to as "extended workforce solutions." At twenty years old, Chang was enamored with the pitch to work at Google on a secretive project. They jumped at the chance to start their career with such a prestigious company. "I remember my friends and family being so proud of me," Chang told TechEquity researchers. "I didn't even know what I was doing yet but [I thought] I should be proud of myself because I get to put Google on my résumé." Chang's expectations were quickly shattered as they came to understand what their role really entailed.

When Chang showed up for their first day of work, at a makeshift warehouse off the back of a Costco, the crop of new workers in their orientation group were told that they couldn't tell people they worked at Google. They had to represent themselves as contractors through Adecco who were placed at Google. Chang thought to themself, "This isn't what I envisioned working at Google was like. Where's the slide? Where's all the free food?" The job was doing pack-and-ship for Google Shopping Express, Google's attempt to compete with Amazon Prime, which was hardly the sexy office job the Adecco recruiters had made it out to be.

Chang's experience working as a contractor at Google is representative of the ways contracting out is undermining labor standards for a growing number of workers across the industry. Like Chang, many workers take these temporary positions because of the opportunity it presents to transition into a great job at a prestigious company. But in reality, those pathways do not exist. Seventy-seven percent of the contractors who spoke with TechEquity never converted to a full-time position. Of those who did, most of the time the conversion to a full-time position happened at a different company than the one where they were temping. The contracting ecosystem often makes it *harder* for temps to land full-time roles than if they'd never taken the job in the first place. Temps who represent themselves as working for the tech company, as opposed to the staffing agency who hired them, are often blackballed. Recruiters interviewed for TechEquity's Contract Worker Disparity Project confirmed that contractors who listed the tech company as their direct employer on their résumé were passed over for future opportunities.

The contract that workers agree to when they start these jobs are for very short increments (usually starting at six months) and are typically renewed at the client company's discretion every month or two until they hit the two-year mark. After two years, the workers are forced to leave for at least six months before they are allowed to start the two-year cycle all over again—a practice that stems from the playbook established in the Microsoft permatemps settlement and allows client companies to maintain the fiction that these workers are only filling temporary roles. Chang was forced to go back to their lower-paying retail job for six months once their initial two-year stint was over. After six months, they were hired back by Adecco into another role at Google Shopping Express.

For anybody trying to plan a family budget, such unpredictable arrangements are hard to manage. Aside from the month-to-month uncertainty the structure creates, it prevents workers from creating stable lives. Most landlords require proof of consistent employment, which means some guarantee that a tenant will be employed through the length of the lease. Mortgage providers require proof of stable employment to qualify a homebuyer for a loan. When contracts are extended, it is often without the contractor knowing about it, denying them the opportunity to try and negotiate for better compensation. Contracts can also be terminated on a whim and at the sole discretion of the client company. Chang's second contract was cut short abruptly after missing a few days due to illness. When they questioned the decision, their hiring manager for Adecco and an HR rep for Google made it clear there was no recourse; Google was free to decide not to renew their contract at any point and for any reason, without any opportunity for Chang to object.

Beyond the logistical complications these short-term contracts present, they also disincentivize workers from speaking up about unsafe, unethical, or inappropriate behavior in the workplace. Eventually, after gaining experience at another company, Chang ended up in a contract creative design position at Google Brand Studio, Google's in-house advertising team. They were making $50 per hour and finally felt they had landed the prestigious tech job they'd been after since the day they met the Adecco recruiter at their campus job fair. Things were going well, but when the protests following George Floyd's killing started in 2020, Chang began speaking up more often about what they viewed as racial inequities in their department, creating tension with their white manager. Not long after, Chang was informed that their contract wasn't going to be renewed because of COVID-19 budget cuts (Google posted a 14 percent revenue increase in the third quarter of 2020, the same quarter Chang was let go). It is of course impossible to know the real reason Chang's contract wasn't extended. But the perception that it stemmed from tensions created by speaking up about workplace issues reflects the tough decisions contractors face; most choose to stay silent. Temp workers know the parent company can cancel their contract at any point for no reason, so better not to rock the boat.

Temp and contract workers who are hired through third-party agencies also face significant challenges when it comes to management and supervision. Their day-to-day work is overseen by a full-time employee of the tech company, but their pay, benefits, and legal employment status are handled by the staffing agency who hired them. At every juncture in Chang's contractor journey, this fragmented management structure played a role in creating or exacerbating harm. When Chang started work at Google Brand Studio, their Google manager was so impressed by their work that they offered a $10 per hour raise. But because they weren't Chang's legal employer, Google didn't have the authority to extend a pay increase.

When Chang's Google manager advocated on their behalf to their legal employer, a staffing agency called Synergist, their manager was told Chang's pay was capped based on the contract terms Synergist had with Google. The only way for Chang to get a raise, they were told, was for another Synergist contractor at Google to get a pay cut.

The fragmented management structure also makes it hard for temps and contractors to perform their roles well. Because the tech companies need to avoid being considered "joint employers" with co-equal legal responsibility for the workers, they create firewalls for temps and contractors that add friction and opacity to their roles. Temps and contractors typically are not allowed to attend all-hands meetings or participate in other internal communications channels where important information is shared. They can't attend staff retreats or holiday parties where team camaraderie is built and networking happens. They can't access internal resources like manuals and handbooks they need to facilitate their work. Their tech company managers are not allowed to offer anything that appears to be a professional development opportunity, including offering any formalized feedback on work performance. This hands-off approach to management, driven by an overriding imperative to avoid legal obligations to the workers, contributes to a sense among workers that temp and contract roles are dead-end positions. In TechEquity's survey of temp workers, 70 percent reported that a raise was unlikely or very unlikely; 80 percent said the same about the prospects of a promotion.

The temps and contractors who fill these roles are very often paid significantly less than their full-time counterparts who do the same work. Seventy-five percent of the software engineers who are hired through staffing agencies, for example, earn salaries in the 25th percentile for all software jobs. Normally, this would be a violation of equal pay laws. But since the workers are not technically employed by the same entity (the temps and contractors are employed by the staffing agency partner) equal pay laws do not apply.

The difference in compensation is even more stark when benefits like health insurance and paid leave are taken into account. Chang was making $16 per hour when they started at Google Shopping Express and received bare-bones health insurance through Covered California, California's state-run health insurance exchange. Their wages increased to $21 per hour in their new role after the six-month break, but the benefits were still significantly less than what Google's directly employed workers received. Google and other large tech employers are well known for their generous benefits packages: several months of fully paid parental leave, 100 percent health coverage for entire families, unlimited vacation and sick time. Chang and their coworkers did not even receive paid sick leave until 2020. When Google announced they would be providing paid wellness days—days when the whole company shut down to take a collective breath—to its full-time staff during the pandemic, Google's contractors had to take those days as unpaid time off. Rachael Sawyer, a contract worker for Google in Austin, told TechEquity she and her fellow contractors received only ten paid days off per year, including all sick days, holidays, and vacation time. They recently fought back to increase that number to fifteen. For holidays like Christmas and Thanksgiving, which Google's full-time employees receive as paid holidays, contractors must use one of their allotted paid time off days. The survey conducted by Alphabet Workers Union found that nearly half of the workers who responded did not have access to paid sick days. Over 70 percent said they did not have access to paid parental leave.

As the tech industry entered a period of mass layoffs in 2022, contract workers were hit hard. Because they are employed through third parties, in most cases they do not receive protections, advance notice, or severance packages when mass layoffs occur. One contractor at Twitter recounted that she found out she was laid off in 2023 only after she found herself unable to access her email. Some tech workers are being laid off only to hear from staffing agencies that are offering them their job back at the same company—but as a contractor instead of a full-time employee.

The unfairness of this dual-track workforce is compounded by racial and gender discrimination. It turns out that the tech workforce is much more diverse than publicly available demographic data about the industry makes it seem. Workers who are from underrepresented minority groups are more likely to be one of these temps or contractors working in the shadows than they are to be full-time employees of the tech companies. According to TechEquity's research, Black and Hispanic workers are 40 and 50 percent more likely to be a contractor in tech than a full-time employee. Most of Chang's coworkers at Google Shopping Express were other people of color, many of whom did not have the same educational credentials that Chang had. On the occasions they got to visit the official Google office, the difference in racial makeup between their team and the full-time employees at Google, who were mostly white, was stark. The racial divide persisted at Google Brand Studio, where all of Chang's coworkers of color were contractors except one Black man. According to data collected by TechEquity and the nonprofit Project Include, members of all gender, racial, and ethnic minority groups are overrepresented in the tech contract workforce compared to their makeup of the workforce overall. Combined with the fact that all racial groups except Asians are underrepresented in the tech workforce, a stark racial divide begins to emerge. By relying heavily on outsourcing, tech companies have created a caste system within their workforce that sees underrepresented minorities in lower-paying and precarious contract roles reporting to an overwhelmingly white management class and with no clear pathways into full-time employment at the parent company.

Taken together, these characteristics of the temp worker experience paint a very different picture of what it's like to work at a tech company. The stereotypical "tech bro" isn't representative of your average tech worker. In reality, a significant number of tech workers receive middling wages and benefits. They have very little stability, protection, or voice on the job, and the pathway to advancement is murky at best.

Because of the lack of transparency around temp workers (companies are not required to disclose data about workers they hire through staffing agencies) and an unwillingness among tech employers to discuss the practice, it is hard to nail down how exactly this phenomenon plays out. From a few interviews conducted by TechEquity with tech company employees responsible for procuring temp workers (none of whom were willing to go on the record), we can gain a rough idea of the internal culture that results in large tech companies employing more temps in office roles than full-time employees.

First, budget allocations for contractors are easier to secure and more flexible than they are for adding full-time head count. Hiring managers described high barriers to adding full-time roles to their teams and said that decisions to add new head count were considered only once per year as part of the budgeting process. Contractors, however, were more fungible, and it was easier to get approval for additional budget to hire through a staffing agency in off-cycle periods.

Second, hiring managers mentioned the agility that hiring temps allowed. Because of the fast pace that venture-backed companies demand (remember the advice from *Blitzscaling* that calls employees growth limiters), being able to add staff capacity to your team on short notice is a must. Filling a full-time role can take months. But hiring through a staffing agency provides almost instant talent since the work of recruiting, vetting, and onboarding is done by the agency.

Yet there were also signs that the practice had creeped beyond just the need to flexibly staff up and down as projects waxed and waned. One hiring manager reported that their company posted roles as if they were full-time positions with the tech company but turned candidates over to staffing or payroll agencies when it came time to hire them, in an obvious attempt to keep these workers off their books.

After taking time out of the workforce to unwind, Chang has finally landed their dream job as a full-time employee doing creative work at another tech startup. "I've noticed the difference between being a contractor and a full-time employee and it's night and day. That fantasy of going from contractor to [full-time employee], it's real. The type of respect you get when they realize you're a full-time person in comparison to a contractor, I notice the difference between how they treat contractors versus how they treat me. I stay close to the contractors and try to let them know I understand what they're going through."

Chang's experience is typical of the tech temp worker experience across the industry. Google isn't alone in outsourcing key functions of the organization to staffing agencies. Investors applaud these companies for keeping their workforces lean, and reward tech companies by pumping up their valuations.

Ultimately the calculus for venture-backed companies to out-source their core staffing needs is the same as it is for gig workers. When you're growing fast and trying to take over markets quickly, adding full-time employees—growth limiters—slows you down. Tech companies have found a convenient way to scale up their staffing capacity without incurring the full costs associated with hiring high-skilled workers. Many companies across sectors hire vendors to provide various services, like security or janitorial work, that supplement the operations of the business. But venture-backed tech companies take the practice further than most, routinely hiring out for functions that are core to their business, including software engineering, marketing, content moderation, and recruiting. Many of the entre-preneurs I spoke to during my research told me (off the record, so as not to upset investors) that they had been pressured to outsource key parts of the business to third parties.

As the tech industry matures and as venture capital continues to expand its investment reach beyond software companies, there will be an increasing need to fill roles beyond those that are core to soft-ware development. But there remains the investor-driven imperative that all venture-backed companies scale as software does, and it will increasingly be in tension with realities of the markets these compa-nies are entering. The tension is particularly stark as the AI sector booms. Venture capitalists are competing hard to throw billions of dollars into the rising crop of AI startups, in the process spurring the incumbent tech giants to scramble to keep pace. The develop-ment of AI, being neither artificial nor intelligent, requires tens of thousands of workers behind the scene to clean and train the data that is fed into AI models, and to manage the outputs that these sys-tems create. AI companies are hiring most of these workers through third-party staffing agencies, often in developing countries with weak labor standards.

The story of Instagram—how the company was sold to Facebook for over $1 billion when it had only thirteen employees—has become lore in Silicon Valley. The standard it set is one that investors, implicitly and explicitly, continue to hold their portfolio companies to, no matter how unrealistic. Contracting out is viewed as a way to achieve an Instagram-style exit, and the workers are the ones who pay the price.

The tech industry has long been plagued by an intractable diversity problem, which can partly be explained by the culture of exclusivity in venture capital. According to a Deloitte survey, conducted in 2020, of 375 venture capital firms across the US, only 16 percent of investment partners were women. Only 3 percent of investment partners were Black, which would be shocking if you didn't know that Deloitte found zero Black or Latino investment partners—*not a single one*—in 2016.

That a branch of American capitalism is heavily white and male is not, unfortunately, abnormal or even novel. But the character and structure of venture capital makes it particularly susceptible to an old-boy-network mode of operating. The role of the venture capitalist is essentially to convince rich people and institutions—the limited partners—to give them massive amounts of money on trust that the venture capitalist can pick the portfolio of companies that will deliver venture-sized returns. As Tom Nicholas points out in *VC: An American History*, the social connections between limited partners and general partners were extremely important in the early days of modern venture capital's evolution. The privacy of the limited partnership structure allowed for an extreme insularity, creating conditions that are rife with bias, exclusion, and discrimination and that have largely gone unchanged as venture capital has matured.

Success also begets success in the venture capital world. LPs want to invest their money with venture capitalists who have a track record of delivering large returns. They compete to be part of the most prestigious VC funds, which in turn allows those VCs to create better terms for themselves. After they hit it big, VCs often turn into limited partners themselves, investing their own money back into venture capital funds. This incestuous cycle makes it very hard for people who weren't already on the inside to break through.

Once women and people of color do get a foot in the door, the toxic culture of the industry does its best to drive them right back out. When Ellen Pao, a venture capitalist at Kleiner Perkins, sued her employer for discrimination in 2012, the veil was pulled back on what it's like to be a woman in venture capital. By Pao's account, she was harassed by men in the office, left out of meetings, and routinely passed over for promotion in favor of men. Her lawsuit became an emblem for women—especially women of color—across Silicon Valley who saw their experiences reflected in Pao's story.

The network-y culture that is the lifeblood of dealmaking between limited partners and venture firms carries over into how those firms invest in startups. In 2013, at the Personal Democracy Forum conference, I gave one of the first public talks about how unequal venture capital funding was. At that time, less than 1 percent of the startups receiving venture capital investment in the US had Black founders. Eighty-seven percent of them had white founders. Despite the attention those shocking figures got, and the community of advocates that emerged to address diversity in venture investing, almost nothing has changed since then. In 2020, Black- and

Latino-founded businesses *combined* made up only 3 percent of all funded startups. As venture capital investment constricted in 2022 and 2023, the situation got even worse. While funding to startups overall dropped by 36 percent in 2022, funding to startups led by Black founders fell by 45 percent.

Unlike other sources of capital investment, such as small business loans, which are extended based on the financial fundamentals of the business, raising venture capital is all about who you know. Since venture capitalists are essentially making bets on whether a founder has what it takes to grow a company to achieve venture-scale returns, investors rely heavily on proxies for competence. Did the founder go to a prestigious university or work at a brand-name consulting firm? Do they run in the same circles I do? "When a pitch crosses my desk, the first thing I do is go on LinkedIn and see who I have in common with the founder," one venture capitalist told me. "If we don't have any connections, it's not a deal breaker. But it's definitely a point in their favor if they know people that I trust."

After a few fund cycles, a pattern began to emerge across Silicon Valley of what "success" in startups looked like. Because the first generation of venture capitalists largely invested in companies run by people they knew, and because those people were largely white and male, subsequent generations of successful startup founders were also predominantly white and male. This fact came to be interpreted as a lesson about who was likely to be a successful startup founder. As Paul Graham, one of the cofounders of Y Combinator, once said about identifying a potential successful founder, "I find myself inadvertently noticing as I'm walking down the street, I see a couple of guys walking down the street and I think 'oh they look like they'd make good founders, or bad founders, as the case may be.'" When pressed to describe the look, he replied, "the right kind of nerdy."

As Ellen Pao told me, this kind of pattern matching quickly becomes a self-fulfilling prophecy. "At Kleiner, they wanted twenty-six-year-old white men who had dropped out of school," she said. "That's who you fund and then, because a prestigious firm like Kleiner Perkins is behind them, that's who succeeds. And then you point to that as a pattern of success. But it's only because that's who you're investing in and supporting and connecting to other resources."

Since the lack of diversity in venture capital has become a headline-making issue, some strides have been made to address it. But the small number of Black and Latino investors who have entered VC over the last decade have faced strong headwinds. There's a "twice as good" thing happening where Black and brown VCs have to overperform the expectations placed on white investors in order to get limited partners to make commitments to their funds.

Some prominent investors have responded to calls to diversify their portfolios by establishing separate funds meant to support minority founders. Andreessen Horowitz established the Talent x Opportunity fund with $2.2 million in personal contributions from its partners, and SoftBank created a $100 million fund for Black and brown founders. The size of these funds, however, pales in comparison to these firms' mainstream investment vehicles. SoftBank's commitment is just a tenth of a percent of the assets in its $100 billion Vision Fund.

Beyond their paltry size, creating these smaller funds that stand apart from the core investment activities of the firm can make the problem of discrimination even worse and perpetuate the idea that investing in minority-led companies is a form of charity rather than a business opportunity. "We should be far beyond separate but equal," Monique Woodward, a Black investor at Cake Ventures, said at an event covered by *Bloomberg Businessweek*. "Black entrepreneurs don't need a separate water fountain. You have to fix the systemic issues in your funds that keep Black founders out."

Black investors also run into challenges when they raise money from limited partners to seed their funds. Charles Hudson, the managing partner at Precursor Ventures and one of the leading Black venture capitalists in the country (he is currently serving as the head of the NVCA), told a reporter at the Bloomberg event, "I meet limited partners and they're like 'oh you're a diversity fund.' Nothing in my deck says we're explicitly prioritizing diversity. They meet you, they see you and they immediately jump to that." I heard another story, from a venture capitalist that prioritizes diversity among their funders, that one limited partner told them they had already spent all their "Black money" for the year so they wouldn't be able to invest. This was despite the fact that this venture capital firm consistently delivered returns that were better than industry average.

The shocking lack of diversity in venture capital investing naturally gives way to an intractable diversity problem at tech companies. Tech is notorious for its lack of diversity among its full-time workforce, a problem that has persisted despite years of public scrutiny and pledges to do better. Black and Hispanic workers make up only 7.5 and 8 percent of the tech workforce respectively, despite each representing about 14 percent of the total US workforce. Women make up only about 36 percent of the tech workforce, despite representing almost half of the US workforce overall. When companies are growing fast enough to satisfy investor demand, they have to hire fast. For the roles that can't be automated or outsourced to staffing agencies, startup executives tend to lean on their networks—and those of their investors—to fill early roles. They also rely on the same pedigree and proxies for success that venture capitalists look for when making an investment, to judge whether people have the competencies to succeed in startups.

Because of the narrow focus on product and growth that investors demand, many startups ignore the culture-building aspects of organizational development until it's too late. HR teams at growing startups run lean and are focused almost entirely on staffing, not things like diversity or developing programs that make for a healthy work environment. One venture-backed startup founder told me that when he asked his investors to help him put together an employee handbook, he was given a version that omitted a paid medical leave policy. When he asked if they could help him develop one (after all, he was an engineer who didn't know anything about benefits programs), he was told he didn't need to worry about it. Instead, he should focus on building the product.

This cultural debt comes back to bite companies when they get bigger. Despite paying considerable lip service to diversity and inclusion, none of the major tech companies has meaningfully increased diversity at their companies since demographic data about the tech workforce became widely available in 2014. Google's Black and Latino workforce, for example, grew from 5.3 and 6.9 percent respectively in 2022 up to 5.6 percent and 7.3 percent in 2023. Most of those increases came in nontechnical roles. These numbers stay stuck in place in large part because the toxic culture at tech companies drives underrepresented minorities out of the industry at higher rates than their white counterparts. In a 2017 report, the Kapor Center for Social Impact found that 25 percent of underrepresented people of color had experienced stereotyping at work, and it was the most common reason for their leaving their employer.

Layoffs swept the tech industry starting in 2022, and since then, there is strong evidence the diversity problem is getting worse. Diversity and inclusion roles were among the most affected by layoffs, and listings for new D&I roles in tech fell by 19 percent in 2022, signaling that tech's interest in creating a diverse workforce was a passing fancy that could be sacrificed as soon as the labor market shifted back in favor of employers.

The tech industry's lack of diversity, from venture capital to the companies themselves, has serious implications for the country's racial wealth gap. Venture capital investment has created trillions in wealth over the last two decades. Facebook, Google, and Apple alone are worth about $6 trillion combined. The vast majority of that new wealth has gone to white men. As a norm, tech companies provide stock-based compensation, which serves to create wealth for their employees in a way that many other industries do not. This is especially true for employees of early-stage startups who, because those startups aren't able to offer salaries and benefits that compete with more established companies, give their early employees a bigger cut of the company itself. These employees essentially trade a hefty salary now for the promise of future riches. Taking this risk requires workers to have a personal safety net that can hold them if the bet doesn't pay off—something many Black and brown workers do not have. As a result, and in concert with other exclusionary hiring practices in venture capital and the tech industry as a whole, they are denied access to one of the greatest wealth-building opportunities of modern times. A report by HR&A Advisors shows that the lack of diversity in the tech industry workforce has led to a $46 billion increase in the country's racial wealth gap.

Labor exploitation is a long and sordid thread running through the history of American capitalism. It manifests insidiously, as it did for Chang and their coworkers as they ran into barrier after barrier on the path to a high-paying job in tech. It plays out obviously in the way workers like Pedro Chevez and Charles Lee find their safety net snatched away days before Christmas. The tech industry certainly didn't pioneer the worst of these practices. But as we have seen from the stories of workers who get the short end of the stick in the tech-driven economy—and who are disproportionately women and people of color—tech companies are mainstreaming and supercharging labor exploitation for a new age. Things like the misclassification Pedro Chevez experienced at Dynamex or the occupational segregation Chang experienced at Google, which were once forms of corporate overreach or greed that happened on the margins of the workforce, are cornerstone elements of venture-backed business models. The returns that investors demand of companies force them to run as lean as possible, and that culture of austerity persists even beyond the point when investors have cashed in. As venture capital becomes the main way that startups get funded, especially AI companies, which pose a threat to workers across the economy, and as other industries eye the venture model for exploiting labor, the potential for labor protections to be eroded even further looms large.

4

VENTURE CAPITAL AND HOUSING

CAPSIZING THE AMERICAN DREAM

I f market size is the key to unlocking venture-scale growth, there is no bigger opportunity than the $43 trillion American residential real estate market—and there is perhaps no other area of the economy as vulnerable to venture capital exploits. Housing is at once the linchpin of middle-class wealth creation and the primary locus for perpetuating inequality in American society. Scholars like Richard Rothstein and Keeanga-Yamahtta Taylor have brought renewed attention to the ways discrimination in the housing market continues to define the difference in life prospects between white Americans and everyone else, especially Black Americans. Redlining in the early twentieth century, predatory home financing schemes like contracts-for-deed in the middle of the century, and the fallout from the Great

Recession in 2008 created a context that is important to understand in order to see clearly how venture capital's entrée into the housing market is undermining our ability to achieve broader economic and racial equity.

The US housing system remains highly segregated, and the problem is getting worse despite a variety of antidiscrimination laws that have taken effect since the civil rights era. One recent UC Berkeley study found that more than eight out of ten American metropolitan areas were more segregated in 2019 than in 1990, and home values and incomes were twice as high in communities that are predominantly white than those that are predominantly communities of color. The legacy of Jim Crow laws still resonates strongly: 83 percent of communities that were redlined in the early twentieth century remain highly segregated by race, and those communities have worse outcomes than predominantly white communities across a sweeping set of indicators, from economic opportunity to public health.

The persistence of racial segregation and exploitation in housing was laid bare during the 2008 financial crisis when the housing bubble burst. Black and Hispanic homebuyers were two to three times more likely than whites to be denied a prime mortgage, but twice as likely to be approved for a subprime mortgage. The distinction was even greater for high-income Black homebuyers, signaling that the effect wasn't simply a function of socioeconomic well-being. As a result, the gap between white homeownership and Black homeownership is larger now than it was when the Fair Housing Act was passed in 1968. In Atlanta—which we'll see is ground zero for venture capital's foray into the residential housing market—more than half of all foreclosed properties were in Black communities, even though the housing stock in those neighborhoods accounts for only 30 percent of the area's total housing stock.

While the 2008 foreclosure crisis was devastating for homeowners, especially homeowners of color, it turned out to be a boon for Wall Street. Even before the dust settled and the taxpayer-funded bailouts had time to take root in the banking system, investors understood that there was a moneymaking opportunity in the rubble of the American Dream.

In the aftermath of the crash, officials at the Department of Housing and Urban Development (HUD) made a fateful decision that laid the groundwork for venture capital to become a major player in the housing market. Six million American families lost their home to foreclosure and at the depths of the crisis, 10,000 properties were being foreclosed on *per day*. In some communities the situation was particularly dire. In Stockton, California, where home values plummeted over 60 percent from 2008 to 2010, one in ten homes was in foreclosure. In 2012, it became the first large city in the US to file for bankruptcy.

HUD officials were understandably alarmed. As the foreclosure wave crested, they feared that many communities would suffer irreparable damage and become full of empty, bank-owned homes that used to belong to families. Part of HUD's response was to create the REO (Real Estate Owned)-to-Rental program, which provided subsidies to Wall Street firms so that they could buy foreclosed single-family homes in bulk and turn them into rental properties. The REO-to-Rental program was designed to address many problems at once. It would pump investment into communities that had been hollowed out by the crisis; provide desperately needed tax revenue to local governments, many of whom were on the brink of bankruptcy; and bolster rental stock, creating more housing options for the millions of people who had turned from homeowner to renter almost overnight.

The REO-to-Rental program enticed Wall Street investors into the single-family real estate market en masse for the first time, and private equity firms came to understand the opportunity it presented. Not only had the foreclosure crisis made it possible to buy homes for much less than they were worth, the exploding population of renters was creating a new market. While homeownership rates flatlined from 2006 to 2017, the share of households headed by renters went up from 31.2 percent to 36.6 percent, creating the largest population of renters since the late 1960s. That combination of undervalued properties and an expanding market of renters created the conditions that private equity investors saw as ripe for exploitation. They could by properties on the cheap (using capital from HUD programs), charge a premium to renters, and realize all the appreciation the homes would gain as the housing market recovered.

But there were still barriers for private equity investors to overcome before they could tap this opportunity. Before the Great Recession, large-scale institutional investors like private equity firms were virtually absent from the single-family market. Wall Street's interest in real estate was mostly confined to multifamily apartment buildings and commercial property. Maintaining large, far-flung portfolios of single-family rental properties was logistically prohibitive. The accepted wisdom was that you couldn't be an effective single-family property investor-landlord unless your portfolio was confined to a specific geography. Without the tools to manage a broader portfolio, the market opportunity wasn't large enough for it to be worth Wall Street's time. Despite the short-term boon created by the REO-to-Rental program, there was still a lot of skepticism that the model could be sustainable. Famed real estate mogul Sam Zell said in 2013, "How can you operate and create scale in that situation? I don't know how anybody can monitor thousands of houses."

He was about to get his answer. The REO-to-Rental program co-incided with two major technological developments—the emergence of mobile technology and the big data revolution—that were in the process of upending myriad other sectors. Some real estate investors recognized that these tools could help them take advantage of the lucrative profit-making opportunity in the single-family market. Two of the earliest were Doug Brien and Colin Wiel, who founded Waypoint, a single-family rental investment property company, in 2011. Based in the San Francisco Bay Area, Waypoint was one of the very first of the companies to emerge in what would come to be called the SFR, or single-family rental, industry. The two chronicled their experience in a 2022 book titled *The Big Long: How Going Big on an Outrageous Idea Transformed the Real Estate Industry*. In it, Brien and Wiel say, "While most everyone was reeling and fearful from the real estate collapse and some investors were looking for Bay Area buy-and-flip opportunities, we looked at the shattered remains of America's housing market and thought we saw a new industry: large-scale SFR ownership and management driven by data and technology . . . SFR wasn't scalable until Waypoint used technology to *make* it scalable."

Brien and Wiel's strategy was to look for foreclosed properties in blue-collar communities populated primarily by immigrants and people of color who had been hardest hit by the crisis. They would buy the foreclosures at a steep discount and then rent them, often to the same types of people who had been displaced from those communities through foreclosure. While unemployment was rising across the country, the Bay Area's job market was relatively insulated from the worst effects of the Great Recession. That resiliency in the job market, coupled with the expanding population of renters, meant rent prices didn't drop much at all in the years following the collapse. Waypoint capitalized on that. They recruited investors and created a fund that would allow them to buy foreclosed properties in bulk, promising a 10 percent annual return. That first fund with outside investors (which, incidentally, was structured much the same way that a traditional venture capital fund is structured) raised $7 million; the Waypoint team deployed that initial capital quickly, raising a second fund shortly thereafter. They soon raised a third and then a fourth fund. By the end of 2012, Waypoint owned thousands of homes and had expanded their holdings beyond the Bay Area to Los Angeles, Atlanta, Phoenix, Chicago, Houston, Miami, and Orlando.

Waypoint wasn't alone in the market for long. Wiel and Brien were proving that, with the right tools, it was in fact possible for single-family homes to become a scalable-asset class unto themselves. Their success gained Wall Street's attention, and soon big Wall Street private equity firms were getting in on the act. Sam Zell, the same real estate mogul who voiced skepticism about large-scale corporate landlordship in 2013, had changed his tune by 2019. "Technology has disrupted the real estate industry," he said. "The single-family rental sector may be a case in which resulting efficiencies have a big impact."

Those efficiencies were largely created at the expense of the people who lived in the Wall Street–owned homes. Francesca Mari, reporting for *The New York Times* in 2020, outlined in painstaking detail the Kafkaesque horror of living in a home owned by a Wall Street landlord. Tenants sign convoluted leases that shift the responsibilities of homeownership—pest control, maintenance, landscaping, even supplying fire and carbon monoxide detectors—onto tenants.

Finding ways to charge hidden fees is a common practice. Tenants are forced to pay rent through online portals that charge convenience fees of 3 percent or more. They even charge fees for delivering notices warning tenants of their delinquency in paying *other* fees. Alana Semuels reported for *The Atlantic* that between the first nine months of 2017 and the first nine months of 2018, Invitation Homes, the largest SFR company in the country since it merged with Wiel and Brien's Waypoint, increased its fee income by 114 percent even though its property holdings had grown only by 71 percent. Perhaps most troubling, the companies have demonstrated a well-documented pattern of blatant disregard for health and safety concerns: pregnant women with asthma have been forced to suffer through 90-plus-degree heat for weeks for lack of a working air conditioner; young children exposed to mold; sewage overflowing into homes.

When tenants try to fight back, Wall Street landlords galvanize their teams of lawyers to overwhelm them, often using their tenants' naivete about the legal system as a weapon. Mari describes the experience of a tenant who lived in a home owned by Waypoint. Waypoint filed an eviction notice against the tenant, only to tell him they were dropping the case and he didn't need to appear in court. That wasn't true; Waypoint showed up to the court date and, due to the tenant's absence, won a victory in pushing the eviction through. (Institutional landlords are much more likely to initiate eviction proceedings than smaller landlords.)

Of course, exploitive landlords have existed for as long as property ownership. But Wall Street ownership, particularly in the single-family market, is harmful in new and unique ways. The distance between these institutional landlords and their tenants leads to an alienation that doesn't exist in traditional rental arrangements where the landlord typically has a name and a face (or at least is geographically closer at hand than Wall Street). Furthermore, unlike in apartment buildings where tenants can band together to advocate for themselves, residents of single-family rentals are isolated, creating a divide-and-conquer dynamic that makes it much easier for landlords to exploit. These structural realities of the SFR market create conditions where renters have almost no agency in their relationship with those who provision their housing.

ENTER VENTURE CAPITAL

Private equity's ability to exploit the SFR market was heavily enabled by venture capital, with Silicon Valley and Wall Street investor motives often playing symbiotic and self-reinforcing roles in the market. As the SFR industry grew, it was becoming clear that the real estate sector, with its combination of huge market value and old-school modes of doing business, presented just the kind of potential for disruption that venture capital loves. Venture investment in the real estate sector was almost nonexistent in 2008 but had exploded by 2017. The value of the worldwide "property tech" market, or Proptech, was over $9 billion that year, more than eighteen times larger than it had been ten years earlier, just before the crash.

While flagship Silicon Valley venture capital firms like Andreessen Horowitz and Sequoia Capital have made significant forays into Proptech, the most active firms have been those established by the Wall Street actors themselves. Fifth Wall, the largest venture capital firm devoted primarily to Proptech, with $3 billion under management, was cofounded by Brad Greiwe. Greiwe is a pioneer of the SFR industry, having also cofounded Invitation Homes, an early forerunner in the SFR sector. Fifth Wall's limited partners include some of the largest residential institutional owners, including Starwood Capital and American Homes 4 Rent.

Part of Fifth Wall's explicit value proposition is that it has the relationships necessary to connect Proptech startups with the largest potential customers in the world. In the firm's 2017 launch announcement, the founders say, "for all of Fifth Wall's portfolio companies, we have structured . . . partnerships and commercial agreements that dramatically accelerate the growth and industry adoption of their products." This creates an interconnected nature of Proptech—where it is impossible to disentangle the influence of traditional Wall Street investors, who invest in the physical assets (the houses themselves), from the venture investors, who invest in the technology that makes it possible to manage those portfolios.

This seamless integration of Wall Street and Silicon Valley in the American housing market is perfectly encapsulated by the trajectory of Brien and Wiel, the SFR pioneers who started Waypoint. By 2016, Waypoint had grown significantly, as had the broader SFR sector. After Waypoint merged with several other SFR companies, Wiel and Brien were ready for a new challenge. They wanted to put their belief in tech's ability to enable large-scale SFR ownership to the test. They started a company called Mynd to realize that potential. Mynd bills itself as an "end-to-end real estate platform that helps investors find, buy, lease, manage and sell residential investment properties." Heavily influenced by the likes of Uber and Airbnb, Wiel and Brien understood that framing their company as a platform that helped other people manage their property holdings rather than as a real estate investment company, as Waypoint had been, they would be able to pitch a story of venture-scale potential to investors. Brien and

Wiel said they would take advantage of the scalability of software to "extract optimal value from assets owned by other people." In other words, they would take a page out of the venture capital playbook and operate as a platform, avoiding the overhead associated with owning a cumbersome portfolio of physical assets. As the SFR market grew, they would achieve a type of flywheel effect: technology would enable corporate landlords to expand their portfolios, and those expanded portfolios would make companies like Mynd more valuable. Each would symbiotically feed on the other's growth, and the limited partners and venture capitalists who enabled it all would cash in on both of the upsides.

While Proptech was still in its nascent stages, Mynd raised an initial $5 million from venture capitalists fairly easily. Mynd grew quickly from there, expanding the number of housing units it was managing by 30 percent every month. This growth clip pleased investors, and Mynd raised their next round of funding just six months after the first. Pete Solvik from Jackson Square Ventures, which led the round, said at the time, "We've been incredibly impressed by Mynd's early progress and rapid growth. The residential property management space is a massive market and ripe for reinvention. We believe Mynd is poised to lead that reinvention by leveraging technology to create efficiencies, offer better service and provide faster insight for real estate investors everywhere." Mynd has since raised over $175 million and, as of this writing, is valued at over $800 million.

Now, Mynd is positioning itself to be the platform on which the institutional SFR market can scale. When I talked to Mynd cofounder Colin Wiel in early 2024, he predicted that the SFR market was on track to match other real estate asset classes, like commercial and multifamily, telling me that within a decade or two he expected institutional owners to hold 40 percent of the market. To take advantage of that expansion, Wiel says Mynd is orienting itself to serve institutional owners rather than small mom-and-pop owners who were Mynd's early customer base. It recently entered into a partnership with Invesco, the behemoth investment firm, to help it purchase 20,000 single-family homes. This is an indicator that the Wall Street–Silicon Valley integration in the housing market is a force with staying power.

IS VC EXACERBATING OR MITIGATING DISCRIMINATION IN HOUSING?

The investment returns generated by the SFR market, and enabled by venture-backed companies like Mynd, are disproportionately made at the expense of the same communities that were devastated by redlining and Jim Crow segregation. One investigation by *The Atlanta Journal-Constitution* found that 80 percent of the census tracts where institutional investors were buying the most homes were predominantly communities of color. While many real estate experts and policymakers dismiss Wall Street's impact on residential real estate because they own such a small share of the country's overall housing stock—less than 5 percent by most accounts—in some neighborhoods, they have effectively achieved monopolies. One predominantly working-class, Black community in suburban Atlanta saw more than half of the homes sold in 2021 go to private equity firms. A Georgia Tech study found that investor activity was responsible for a 1.4 percent drop in homeownership in the Atlanta area—but the drop was almost four times higher for Black residents. White homeownership rates, on the other hand, weren't affected at all.

The investor activity is driving up property values in those neighborhoods faster than they are increasing in white neighborhoods, a sign of gentrification that is slowly eroding Black communities. As Black Americans are simultaneously increasingly excluded from homeownership and displaced from communities where they have deep roots, the racial wealth gap is getting ever wider.

Wiel and Brien are bullish on the future of SFR, and the potential for tech and venture capital to help the industry scale. Even as investor activity in the single-family market slowed as interest rates have gone up, they predict that institutional investors will purchase 300,000 to 400,000 single-family homes per year over the next five years. "This is a trend that is just now beginning," Wiel and Brien say. They weren't deterred by the post-pandemic housing market stagnation. In 2023, with interest rates near their highest point, Brien told an interviewer, "These groups have a mandate. [SFR] is an asset class now and everyone is underexposed. As you look at real estate, single family rental stands out as one of the most compelling [opportunities]. So I think as soon as there is clarity around the cost of debt and neutral to positive leverage can be achieved, you're gonna see a wave of capital come in and it'll look a lot like 2011."

He isn't alone in his optimism. As the Federal Reserve signals that the trajectory of interest rates is moving downward, investors are betting that the lack of housing supply will continue to make SFR a lucrative asset class. MetLife Investment Management estimates that 40 percent of single-family rentals will be owned by institutional investors by 2030. That's almost 10 percent of America's entire single-family housing stock, a level that will result in millions of families missing out on ownership opportunities in the coming decades. "We want to get to one million homes in the next 15 years or so," said the CEO of Main Street Renewal, a private-equity-backed investor landlord that uses algorithms to purchase distressed properties and flip them into rentals. In 2021 alone, institutional investors committed $60 billion to purchase single-family homes.

Even as the housing market has cooled since the white-hot pandemic buying streak, when interest rates were at historic lows, the fundamental challenges that the financialization of the single-family market presents will persist. For decades, home builders who produce the types of entry-level homes that help working-class people get on the property ladder haven't been building enough of those homes to keep up with demand. The country has a deficit of at least four million homes, a gap that has persisted since the recession. Many of the builders who are picking up the pace of development are doing so in order to sell those new homes directly to institutional investors, who are converting whole communities of newly built starter homes into rentals—cutting regular buyers out of the market before they even have a chance to compete. So-called built-for-rent activity skyrocketed in 2022, with as much as 13 percent of new housing starts on single-family homes going directly to the rental market. As these trends continue, as the population of renters continues to grow, and as high interest rates and stagnating home prices push homeownership further out of reach, the appeal of single-family rentals is unlikely to disappear anytime soon.

With too much demand for too few homes, something has to give. In 2023, the national homelessness rate increased by 12 percent over the previous year. That's more than four times higher than any previous year-over-year increase. No longer an issue reserved for expensive megacities, homelessness is now increasing across the country, among all demographic groups. Many others are living on the brink, with two-thirds of the poorest 20 percent of Americans paying more than half of their income to rent and utilities.

The two-headed hydra of Wall Street and Silicon Valley's foray into housing isn't entirely, or even mostly, responsible for this housing crunch. A variety of factors—exclusionary zoning, the desire for larger homes, people forming families later in life—play a role in our country's chronic lack of housing. But institutional investor participation in the market certainly is making homeownership, historically the cornerstone of middle-class wealth creation in America and the central symbol of the American Dream, even less attainable.

THE MYTH OF DEMOCRATIZATION

Naturally, a level of cynicism has settled over the millions of Americans who have been locked out of the homeownership market over the last fifteen years. This cynicism is especially high among young people and in communities of color. A 2022 survey commissioned by Ariel Investments and Charles Schwab found that 28 percent of Black investors didn't trust financial institutions, compared with 18 percent of white investors. A 2023 Deloitte survey recently found that over 60 percent of Millennials and Gen Zs believed that buying a house would become harder or impossible in the near future.

Cynicism and hopelessness have opened the door for financial products that offer alternative pathways to wealth creation. In the early 2020s, venture capital went all in on cryptocurrencies, which were marketed heavily to those who felt alienated from mainstream finance. The 2022 Ariel-Schwab survey found that Black investors were much more likely to be invested in cryptocurrency (25 percent for Black investors compared to 15 percent for white investors) and for crypto to have been their entry point to investing rather than a more traditional (and less risky) mode like stocks or real estate. In an attempt to undercut savvy Wall Street investors, Millennials and Gen Zs led a craze around meme stocks in the early 2020s, pumping up the stock price of companies that many hedge funds were attempting to short.

The popularity of meme stocks and cryptocurrencies among this cohort, and the financial fallout that many of them have faced as a result of the collapse of both, demonstrates the risk of financial products in the market. When stock in GameStop, the seminal meme stock, fell 60 percent in one day in 2021, it took billions of retail investors' money with it. In 2023 alone, $1.7 billion worth of crypto was stolen from owners through various hacks.

Amid this cynicism, venture capital has recognized a market opportunity. In recent years a cottage industry of fractional ownership companies has emerged, promising to open the world of real estate investing to people who can't afford to purchase property on their own. These companies, who often leverage blockchain and cryptocurrency technologies, allow individual investors to purchase small slices of residential rental properties and receive a proportional amount of the income they generate. Where Wall Street investors created portfolios of many individual homes that were securitized and traded, these companies take that logic one step further and allow investors to create portfolios of tiny slices of many different properties.

They market themselves as an avenue for people who don't have the financial capability to own property through traditional means—particularly younger people who have been priced out of the market—to get a small piece of the pie. "The barrier to entry for real estate investing has always been so high," says Jerry Chu, the CEO of Lofty AI, a graduate of Y Combinator. "We don't believe that should be the case." Landa, another of these startups, is "making property owner-ship inclusive, for the first time ever." Roofstock, which was co-founded by former Waypoint executive Gary Beasley (yet another example of the interconnection between Wall Street and Silicon Valley in the housing market), offers "radically accessible real estate investing." You get the idea. It is hard to say how large this subsector is, but there are several dozen fractional ownership companies in operation at the time of this writing that have together raised hundreds of millions of dollars from venture capital investors. As of early 2022, Roofstock alone had 15,000 homes on its platform and had facilitated over $5 billion in transactions.

Their marketing reflects a common trope in the marketing materials of countless venture-backed financial tech companies: that the erosion of middle-class opportunity is actually a blessing in disguise. It's a trope that evolves out of a broader rhetorical theme that has long existed in Silicon Valley, which positions venture-backed startups as forces for democratization that will disrupt old and exclusive ways of doing business to create broad-based opportunity for everyone. What started as a genuinely held belief that the internet could lower barriers to entry to almost every aspect of society has morphed, in the wake of an economic recovery that has left millions of people worse off than they were before the rise of the social web, into a cynical newspeak that capitalizes on the sense of alienation many people feel from the economy.

It all has the whiff of "predatory inclusion," a phrase coined by Princeton scholar Keeanga-Yamahtta Taylor in her 2019 Pulitzer Prize finalist book *Race for Profit*. She describes the way in which Black homebuyers in the post–civil rights era have been exploited by predatory home finance products that purported to extend home-ownership to those who had historically been locked out of the property market. Black neighborhoods that were still suffering the consequences of redlining were deemed "subprime" by mortgage lenders, keeping reputable lenders out and paving the way for the shady mortgage bankers who contributed to the Great Recession. As Taylor puts it, "Though race was no longer a factor [in mortgage lending decisions], its cumulative effect had already marked Black neighborhoods in such ways that still made them distinguishable and vulnerable to new forms of financial manipulation. Inclusion was possible, but on predatory and exploitative terms." With the federal government now willing to insure mortgages in redlined neighbor-hoods where poor families were desperate to become homeowners, these mortgage lenders were happy to make loans to people they knew were at high risk of default, regardless of the impact the de-faults would have on Black families and the communities where they lived. These practices came to a head in the wake of the Great Reces-sion when foreclosure rates were disproportionately high in Black and Hispanic communities.

The entrée of crypto, and products that mimic crypto's ability to fractionalize ownership of real assets, into the real estate market—and the ramifications that has for retail investors—has raised alarm bells for legal scholars. One paper by legal scholars at the University of Missouri, the University of Iowa, and the University of Pennsylvania points out that when real estate is transacted through crypto tokens, the normal legal rights afforded through property law are upended. This puts real estate transactions that are facilitated through blockchain technologies into an extremely murky legal gray area that even well-seasoned legal professionals are still working to clarify.

The risks of attracting unsophisticated investors into these fractional ownership schemes is obvious even to some venture capitalists who specialize in Proptech. Clelia Warburg Peters, the scion of a high-end real estate brokerage firm who has been investing in Proptech companies since 2015, has said "the level of education in the public is pretty low . . . This is literally exposure to a direct asset or a pooled vehicle of private assets, there is both education needed and risk." Demonstrating the risks, one Lofty AI user posted an analysis of their experience investing on the platform over the course of two years in the early 2020s, showing that, of the fourteen properties in which they had a stake, only two of them delivered the returns that Lofty promised and two-thirds of them didn't deliver any returns at all, remaining unrented over the term of the investment. Ultimately, this user lost money, despite Lofty's claims that investing in rental properties was a safe bet. This user concluded, "I wish I did not invest anything in Lofty. I definitely share much if not most of the blame, as I am obviously unqualified for evaluating these properties as rental properties. But I do have to wonder what's going on on Lofty's end since they seem unable to rent out ⅔ of my properties."

There are also substantial risks for the tenants who live in properties owned by companies like Lofty. If the conditions of living in properties owned by absent Wall Street landlords are as bad as has been widely documented, living in a property that is owned by hundreds or thousands of different people who may be scattered around the world seems particularly unappealing. While day-to-day management is left up to local property management firms, the extreme alienation of tenant from landlord exposes renters to even more risk of mistreatment. Lofty, which operates on the blockchain using a cryptocurrency called Algorand, employs what is called a Decentralized Autonomous Organization (DAO) to oversee its properties. A DAO, which is a popular governance system among crypto enthusiasts, allows communities of crypto token holders to democratically make decisions about the projects in which they are invested. In Lofty's case, each investor who owns tokens representing a share of a property receives a voting stake in the DAO that is proportional to the value of the tokens (no single investor can own more than a 15 percent share of a property). The DAO is tasked with making some decisions about the property that would normally fall to a landlord or property management company: whether to serve eviction notices, how to handle maintenance requests, and how to set rents, among other things. Token holders are also allowed to propose their own votes. For example, according to a 2022 NBC News report, one DAO representing the owners of a Lofty property in Memphis were tasked with approving a tenant's request to fix a broken ceiling fan. The fix was approved, but not until there had been significant back-and-forth among token holders about whether it was really necessary.

According to Lofty's rules, decisions need a 60 percent vote from the DAO in order to be approved. While Lofty claims that the DAOs are not allowed to vote on anything that would violate the law—those decisions are purportedly left to the property managers—it isn't clear that anyone at Lofty is ensuring that tenants' rights are in fact protected.

If there is a violation for which the landlords are found to be liable, the responsibility falls squarely on the investors, not Lofty. I spoke to several lawyers about Lofty's model, some who specialize in landlord-tenant law and some who are experts in corporate governance. While it is impossible to speculate on the precise legalities of Lofty's business model, what these legal experts do agree on is that the company and its customers are treading on very uncharted territory. Based on what it is possible to discern from Lofty's publicly available materials, the investors in Lofty's properties have been shielded from personal liability since the properties are held in LLCs. But if there was a legal judgment against the LLC, there could be fines levied that would eat into the investors' expected earnings. In the worst-case scenario, they could even end up losing the property altogether.

THE PERILS OF ALTERNATIVE PATHS TO HOMEOWNERSHIP

While fractional ownership schemes capitalize on the vanishing American Dream by playing on customer cynicism, other models tap the combination of hope and desperation that many who have been pushed off the bottom rung of the homeownership ladder are feeling. One particularly popular emerging business model is known as rent-to-own (RTO), a form of owner financing in which a company buys a house on behalf of someone who wants to be a homeowner, but can't qualify for a traditional mortgage, and leases it back to them. These arrangements are designed to last for a few years until the prospective homebuyer is able to purchase the house for a pre-agreed price that is set at the beginning of the lease term.

Rent-to-own isn't a particularly innovative business model. It has its roots in a predatory model that was popular in the pre–civil rights era. This model was alternatively called contracts-for-deed (CFD), lease-purchase, or land contracts. They each had specific nuances but shared the same basic structure. Black homebuyers who were locked out of the traditional mortgage market because of redlining had few other options to access homeownership than to enter into these predatory arrangements. In a CFD arrangement, which closely matches the modern-day RTO models, the seller would self-finance the purchase, allowing the buyer to move into the house and make monthly installment payments over the course of a contract term, which could be as long as thirty years.

Unlike with a traditional mortgage, the seller retained ownership of the house until the house was completely paid off and all the terms of the contract were met, but the contracts were usually structured to be impossible for the renters to fulfill. They could not be even one day late with a single payment, they were responsible for all taxes and insurance, they were to perform all maintenance on the property, and the contracts often came with large balloon payments at the end of the term which were almost impossible for the buyers to make. If the terms of the agreement were not kept, the seller could evict the tenant at any point without returning any of the money the tenant had paid during the period of the contract. Oftentimes, the seller would then find another unsuspecting buyer for the same property and run the scheme all over again. Ta-Nehisi Coates, writing for *The Atlantic*, said of CFDs that they "combined all the responsibilities of homeownership with all the disadvantages of renting."

This practice was one of the major ways that Black families were stripped of wealth in the twentieth century. One estimate puts the total cost to Black homebuyers in Chicago in the 1950s and 1960s, 85 percent of whom purchased their home with a CFD, at $3.2 to $4 billion in 2019 dollars.

After the Fair Housing Act was passed in the late 1960s, the most predatory aspects of CFD became obsolete, since Black homebuyers now had (at least in theory) access to more mainstream financing options. But the basic model persists today as RTO. Research from Pew Charitable Trusts estimates that ten million Americans have used an RTO product, or something like it, to purchase a house. These arrangements are most common in distressed communities in the Midwest and South and disproportionately in communities of color. In 2015 in Detroit, for example, more homes were purchased with RTO-like contracts than with traditional mortgages. Some of these products are peddled by fly-by-night local real estate investors. Some are sold through more established companies. And, increasingly, some are venture-backed startups that bring a more modern, digital aesthetic to the practice.

The emergence of venture-backed RTO companies is in many ways the natural evolution of Silicon Valley's foray into real estate, combining elements of both the Wall Streetification of the landlord and the democratization of ownership tropes. Heavy investment in the tech that enabled Wall Street owners to expand their single-family empires bred an understanding among venture capitalists about how valuable it could be to own the underlying assets in addition to the technology that managed them. RTO not only presented this opportunity; it also was a model that made it easy to spin a tale about how technology could democratize access to the housing market.

The first, and by far the largest, of these venture-backed RTO companies is Divvy Homes, founded in 2017. Divvy was soon followed in 2018 by Landis and in 2021 by Pathway Homes, a subsidiary of Invitation Homes, one of the largest Wall Street SFR companies. All three have had investment rounds led by some of the most prolific VC firms in the world: Divvy from Andreessen Horowitz, Landis from Sequoia, and Pathway from Fifth Wall. All three heavily rely on the marketing pitch that their products make it possible for people who have been locked out of traditional pathways to homeownership to get on the property ladder. All three companies have grown quickly, justifying new rounds of investment from venture capitalists. Divvy, which has raised over $1 billion (including debt, which it uses to purchase the properties) and is valued at almost $2 billion, owns over 7,000 homes in 19 markets in the South and Midwest as of this writing. The younger Pathway and Landis have each raised about a quarter of a billion as of this writing. As a sector, they are on a path to potentially rival the size and scale of the Wall Street single-family investors.

Their offerings sound great on paper. The companies give approved customers a shopping budget to buy whatever house they want, with a few constraints, within approved markets. They then purchase the home and let the customer move in for a small down payment (usually 1 or 2 percent of the purchase price). The customers agree that they will purchase the house for a set amount, the "strike price," within a set period of time, usually around three years. That strike price, which is already set higher than the value of the property at the time of purchase, increases by 3 to 6 percent per year, regardless of how the market is performing. In an example reported in *The New York Times* in 2023, Divvy purchased a home for one Atlanta-area homebuyer for $284,000 and set the strike price at $347,000. That meant, at the end of three years (Divvy's typical term length), and assuming a conservative 3 percent annual increase in the strike price, the buyer would be contractually obligated to purchase the home from Divvy for almost $380,000—a guaranteed increase in asset value for Divvy of almost 35 percent. That is an incredible yield for a real estate investment, one that is guaranteed not by market forces but by the desperation of people who are eager to get one foot on the ladder to the financial stability afforded by homeownership.

In addition to the financial risk they create for their customers, there is substantial evidence that these RTO companies perpetuate the same health and safety issues that private equity owners have inflicted on their tenants. Reporting in both *Fast Company* in 2022 and *The New York Times* in 2023 highlights multiple instances of Divvy's unresponsiveness to serious maintenance issues that should have been uncovered during the home purchase process but somehow got overlooked. In the example I mentioned above, where an Atlanta-area customer was locked into a purchase price more than 20 percent higher than the value of the home, "rainwater often seeped in. The electrical system was faulty. Some appliances didn't work. And mold was spreading on some walls." *Fast Company* reported on the experience of a family renting through Divvy in Arizona who were left without working heat in the middle of winter for weeks.

> *"Our responsibility in this situation is to ensure that the mechanical functions of the system are operating as designed," a Divvy regional construction manager told [the tenant] in an email in early February.*

He suggested that [the tenant] "adjust the vents (free option)" or "install an electronic damper system (at your cost)." She could also buy space heaters, he added.

These companies all came of age during a housing boom, when constricted supply and low interest rates made it appear that house values would appreciate at a fast clip forever. Home values had been on a consistent growth path, increasing almost 60 percent since the post-Recession nadir in 2010. In those conditions, it is easy for a prospective buyer to believe that their home will actually be worth what they agreed to pay for it once it comes time for them to exercise their purchase option at the end of their three-ish-year lease term. That Atlanta buyer's home may very well have been worth 35 percent more than Divvy paid for it if they entered into the agreement in early 2019 and transitioned to homeownership in 2022.

But what happens when the market goes down, or stagnates, as it has in many areas since 2022? What if, when a buyer is ready to purchase their house, it is worth less than the amount they are contractually obligated to buy it for? It will be almost impossible for the buyer to get a mortgage for more than the value of the property, meaning they will either have to come up with the difference in cash to buy out the RTO company—or they will be forced to walk away, leaving all their equity behind.

The companies themselves have not given any indication that they are prepared to rework how they calculate strike prices. While all the companies would argue that it is in their financial interest to work with customers to keep them in the properties, the numbers belie that claim. In the Atlanta area, where Divvy owned just over a thousand properties by the beginning of 2023, the company had filed eviction notices on 190 of them, making up almost 20 percent of its customers in the area.

Evicting these tenants may actually be more lucrative than selling them the home. The companies don't make money only when tenants successfully purchase the home. If the renters fail to convert to homeownership for whatever reason—they can't qualify for the mortgage or they decide they want to move—Divvy and Landis, in addition to maintaining the house as an asset on their books, require the tenants to pay a move-out fee equivalent to 2 to 3 percent of the purchase price. Pathway doesn't charge its tenants a move-out fee, but it does increase the monthly rent price that it charges each customer by several percent every year. After the tenants are out, the companies are free to flip the property to a new customer and start the process all over again.

Because of the opacity of these models, it is very hard to know how the economics play out for the companies. RTO and other alternative forms of home finance are not subject to the same stringent reporting and consumer protection laws that mortgage providers must follow. But there are clues. Perhaps the most telling is how few potential buyers convert to homeownership. Divvy, the only venture-backed company that makes its conversion rates public, sits at just under 50 percent. One Divvy data analyst who I spoke to said the company assumes that rate will go down over time, as the customer base grows and more of them complete their lease terms.

For a product that purports to be about making homeownership accessible to everyone, it is very hard to see how converting fewer than half of its customers can be considered success, especially when that number goes down as the company expands its reach. Which makes you wonder: maybe converting people to homeownership isn't what success looks like for these companies? After all, the RTO companies profit from both rental payments and asset appreciation all while convincing their customers to do the basic maintenance and upkeep of the homes. If CFD agreements represented all the responsibilities of homeownership with all the disadvantages of renting, then for RTO companies like Divvy, the model represents all the benefits of being a landlord without any of the responsibilities of property ownership.

Divvy's CEO, Adena Hefets, speaking to an investor forum hosted by Andreessen Horowitz in 2020, made clear that the aspirational homeownership pitch Divvy makes to customers functions as a tool to help them increase their bottom line. "When your customers think and act like homeowners," she said, "you have lower vacancies, lower turnover, and lower maintenance costs per home." She made this comment while showing the audience data reflecting that Divvy was able to charge higher rents and spend less on maintenance than traditional SFR landlords. On this point, she is in lockstep with her investors. Alex Rampell, who leads real estate investments for Andreessen Horowitz and who led Divvy's Series A funding round—and sits on the company's board—echoed the value of selling the aspiration of homeownership to Divvy's customers. "Nobody takes good care of a rental," he told an investor conference in 2018. "But if you're planning on living there at the end of three years, you're going to take better care of it."

Is there really any difference, then, between the venture-backed RTO companies and the Wall Street SFR landlords, other than that the venture-backed companies seem to have found a way to extract yet more profit from their customers? Even Hefets has indicated that the company is inching away from its mandate to move people into homeownership, though it's clearly important to keep dangling the prospect of homeownership at some ill-defined future period in order to make the sales pitch work. "We're going to be moving toward more of an evergreen [model] where people can just continue to reup, reup, reup, so that they always constantly have that opportunity [to buy]," Hefets told *Fast Company* in 2022. "Just stay in that house, just save up more. It won't practically, on the deed, be yours, but you can still paint the walls, you still live in it, you still build equity, you still get the benefits."

In fairness to Hefets and Divvy, in the economic context in which she was making these statements, it made a lot more financial sense for people to continue renting than to enter into a mortgage when interest rates are at their peak and home values may have dropped. But given the likelihood that most of these renters couldn't afford a mortgage in the higher-interest-rate environment anyway, it isn't exactly an act of charity for Divvy to keep them in their homes paying rent and tending the embers of their dimming hopes of homeownership.

Given the lack of transparency and oversight in the RTO market, combined with the fact that startups, as private companies, rarely report comprehensive data about their products, it is hard to say how representative these anecdotes are of the market overall. Divvy is the largest of the venture-backed models and it has spoken much more about its work on the record, which makes it easier to shine a light on its practices. It is probable that Landis and Pathway operate much the same way, but because venture-backed companies are allowed to operate in almost complete obscurity, it is impossible to know for sure.

Without any of this information, it's also hard to know how stable these companies are. Could they go out of business? It's a prospect that is distinctly possible given the current state of the housing market and the extra risk venture capitalists attach to their portfolio companies. Divvy and the other companies in its cohort have leveraged their portfolios of single-family homes to raise massive amounts of debt in order to grow their portfolios even further. They must make those debt payments somehow. If they can't . . . ? Presumably, they could sell their portfolio of single-family homes to another institutional buyer. When Zillow decided to get out of the business of buying single-family homes, it sold the thousands of properties on its books to other institutional buyers. The added complication with RTOs, however, is that unlike Zillow's properties, these homes have renters living in them. Unlike when a mortgage gets sold, there is no guarantee that the new owner will honor the RTO clients' agreements, and the regulatory structure around these financial products is so fragmented and murky that it is unclear whether the clients would have any recourse to stay if the new owners wanted them out. I asked the companies for clarification. Divvy's Hefets responded by pointing me to the FAQ section of their website, which doesn't contain any detail about what happens to customers if the company goes out of business. Landis and Pathway did not respond at all.

On balance, it's hard to see who these venture-backed RTO models help (aside from the companies themselves, of course). To be sure, there are happy customers who, without these products, wouldn't have been able to buy a home. But, even if they were able to lock in a strike price that gave them instant equity when they closed, they still will have spent a significant period of time paying a premium for rent (rents for RTO are consistently far above comparable market rents) as well as bearing other maintenance and upkeep costs that would have otherwise fallen to a landlord. If conversion rates were higher, it would be easier to stomach the idea of people paying a premium to get on the homeownership ladder, but the abysmally low frequency with which people end up purchasing their home makes

it hard to view these companies as anything more than SFR companies who found a way to juice profits by appealing to lasting belief in
the American Dream.

T
he impact of venture capital in how companies like Divvy go
about their business can be hard to pinpoint. But it becomes
clearer when we look at other kinds of RTO companies whose business models don't rely on venture capital—and realize that they do a
much better job of moving people into homeownership. One of those
companies, Trio Residential, which manages to convert about 80
percent of its customers to homeownership, has developed a model
that aligns business interests with the best interests of its customers.
Trio partners with government entities to provide mortgages to
prospective buyers in that entity's jurisdiction. Trio originates a
mortgage that is held by the government partner on behalf of the
prospective buyer during the three-year lease period. When the prospective buyer is ready to transition to homeownership, the mortgage is simply transferred, with all of its original terms intact, from
the government partner to the buyer. Since no new mortgage is being generated, there is no need to reappraise the property, thus protecting the buyer from any fluctuations in the real estate market.
The buyers also have the certainty of knowing what their interest
rate and monthly payments will be, and they get to benefit from
all the equity that has accrued in the property as they have been
renting it.

Trio's model is profitable. The company makes money in a way
similar to normal mortgage providers and servicers: through origi-
nation fees when the mortgage is issued and the lease is signed, and
through a monthly fee that covers their administration of the mort-

gage and the services they provide to the buyer. Unlike the venture-backed models, their strike prices increase by, at most, only 1 percent per year. For families whose income is less than 80 percent of the national median, the strike price is even lower.

Trio's founder, Darryl Lewis, is a veteran housing practitioner with a background in accounting and technology. He started developing the Trio model in 2001, after watching his neighbors in Seattle struggle to get into homeownership in a rapidly appreciating market. Lewis founded Trio as a nonprofit, but he soon found that the economics of his business were a better fit for a for-profit model. When Lewis was starting Trio, the idea of setting it up as a venture-backed startup wasn't a realistic possibility given the state of the economy in 2001. Through a lot of guile and creativity, he landed on the public-private partnership model under which Trio operates.

Admittedly, according to Lewis, that model comes with a lot of overhead. He and his partners had to jump through a lot of hoops with HUD and other government agencies to make it work. Its complexity makes it hard to scale, and even harder for others to replicate. Ultimately, though, it does work, and not just for the company but importantly for the buyers who have a path to homeownership that is only slightly more expensive—and significantly less risky—than the traditional path.

Unfortunately, the model that Divvy and other venture-backed companies have elected is the path of least resistance, from a capital perspective. Lewis is reluctant to criticize any of the RTO firms. He understands that market dynamics make it almost inevitable that these models will take the most market share. But he has seen it close up enough to have a clear point of view about what these products fundamentally are, when you peel back the marketing. "It's very simple," he told me. "It's a more efficient SFR structure. They sell some [houses], they keep the majority, they get someone to pay premium rents and then sell them later or get someone to reoccupy them at better rates."

This is what makes companies like Divvy so attractive to venture capitalists. They use a message of democratization of access to juice the already lucrative returns Wall Street investors were making on the SFR market. In doing so, they have found a way to tap into the American homebuyer's desperation and undying hope in the American Dream. In the process, they not only rake in even more returns for investors, they crowd out more sustainable models like Trio that work on a smaller scale—much the same way Shef did to Foodnome—and that have better alignment with positive societal outcomes.

———

This chapter has been the most difficult for me to write. Disentangling the influence of venture capital on the housing market from the other negative forces at play, in an area of the economy that is shot through with inequities centuries in the making, is a nearly impossible task. Venture capital may have enabled private equity's foray into the single-family market, or it may have just accelerated the inevitable. The Silicon Valley money supplied to fractional ownership companies may provide the lifeblood for a particularly risky investment opportunity, but the cynicism and alienation that create a market for those companies stem from the rising force of inequality that has been growing for decades. Venture capital is enabling the scale of RTO models that seem predatory in almost every application, even when it is clear there are RTO models that do align business interests with the best interests of potential homeowners. But these models are not the exclusive purview of venture-backed companies. The largest RTO company, Home Partners for America, is owned by the Wall Street private equity firm Blackstone. If venture capital left the market, it isn't clear how much of these harms would be averted.

In fact, while venture capital's participation in the housing market is clearly causing some harm, I believe the problem of venture capital in the housing market might be that there isn't enough of the right type of venture capital rather than that there is too much. The housing market is a stark example of an area where innovation is desperately needed, but the business models that can make a positive difference don't align with the kind of returns venture capitalists and limited partners expect—and so they don't get funded. Much as with Shef and Foodnome, the flood of venture money to companies like Roofstock and Divvy drives out promising but less lucrative ideas, forcing entrepreneurs either to adopt a potentially harmful venture-style approach or to limp along without the investment they need to thrive.

Alex Lofton's experience running a venture-backed Proptech company called Landed highlights the challenges faced by well-meaning housing innovators as they try to grow for-profit solutions to the housing crisis. I met Lofton in 2012 when I moved to San Francisco to open the Obama campaign's tech office. He was a field organizer for President Obama's first campaign in 2008. Afterward, flush with the techno-utopianism that working on an Obama campaign tended to inspire, he moved to San Francisco to work at a startup building political technology. He was excited by the heady Silicon Valley ethos of the time and enrolled in Stanford's Graduate School of Business in 2013 to gain the skills he thought he needed to start his own company. It was there that he committed to the prospect of building a company that could help people access homeownership. It was important for Lofton, a Conflicted Altruist through and through, to be working on something he felt had deep moral purpose. For a kid whose parents, a teacher and a social worker in Seattle, were able to access middle-class stability because of a house his grandmother gifted to his mother, it troubled him that homeownership seemed so far out of reach for his generation. Lofton spent the last six months of his time in business school developing a model that allowed prospective homebuyers to gain access to down-payment assistance in exchange for a piece of the home's equity, and he started Landed to bring it to market.

Landed launched in 2015, initially partnering with school districts to provide down-payment assistance to district employees so they could own a home in the communities where they worked. It later expanded its offering to all essential workers, like nurses and firefighters, whose ability to put down roots close to where they worked contributed to social cohesion. Lofton was initially hesitant to take venture capital, but when he evaluated what appeared to him to be his other options—either to set up as a nonprofit or to bootstrap with his own capital and the money the business could bring in on its own—they weren't the right fit for his ambitions. When I asked Lofton what it was about venture capital that made him so reluctant, he likened it to getting on a runaway train that was constantly ratcheting up the speed. "I'm all for meeting high expectations," he said. "But I don't want to be out of reality. I want to be able to respond to the economy, take the time to build the right kind of relationships and not disrupt the communities in a bad way."

Lofton secured initial funding from friends and family, hoping to have a working business model established by the time that money ran out. But that didn't work out and he realized he'd need to seek funding from institutional investors to take the company further. He decided to apply to Y Combinator and was accepted into the Winter 2016 class. "The conclusion was that in order to make this work we needed to move quickly," Lofton said. "We needed a lot of money up front to get enough customers that we wouldn't be a niche product. Venture capital was the way to go." Lofton admits that part of the reason to take venture capital was the cachet it imbued on the company and on him. But he also didn't think he had very many other options. "Living in San Francisco at that time, raising venture capital is just what you did," he said. "It was part of the culture."

Once he decided that venture was the path forward, Lofton found ways to rationalize the decision, hoping he and his team could thread the needle that would allow them to get off the venture treadmill without giving up too much control or sacrificing their mission in the process. "We did always know that if we couldn't meet our goal of raising money for the down payment program that was necessary to scale our product, there was a fundamental mismatch between what venture needs and what we could become," Lofton told me. But Lofton hoped he could achieve the pipe dream that many other entrepreneurs also expressed to me: being able to raise venture capital and buy out investors quickly before giving up too much control of the company. "Even if venture wasn't a perfect fit long term, we thought we could take advantage of the moment to inject enough capital into our business to get big fast enough so that we could stand on our own while returning capital to investors. That was the dream." To get there, they needed to sell venture capitalists on the idea that they could become a unicorn, whether that was realistic or not. "We started with the billion-dollar outcome and worked backwards to the business model that we thought would get us there," he said.

In the years after his time at Y Combinator, it seemed like he might be able to pull that off. Landed grew quickly, demonstrating strong enough growth to raise a Series A funding round and, shortly thereafter, a Series B. At Landed's height, it was licensed to operate across most of the country and had facilitated down payments for 1,500 essential workers (mostly teachers) to purchase almost a billion dollars' worth of property.

But when interest rates began rising in 2022 before Landed had reached that elusive point of sustainability, things turned sour quickly. They were no longer able to raise capital for their down-payment assistance fund, prompting a series of hard conversations between Lofton and Landed's venture investors. That's when Lofton really started to understand the consequences of building a business on a foundation of venture capital. "There was a moment when the conversation stopped being about how do you make this down payment product successful and it became more about how do you pivot to a profitable business model," he said. Instead of providing down-payment assistance to essential workers so they could put down roots in their communities, the Landed team was urged to explore other more profitable models. Where some venture-backed entrepreneurs get excited about building a successful business, no matter the product, Lofton's heart wasn't in it to chase profit for profit's sake. He was passionate about Landed's mission and didn't want to run a company that was leaving that behind.

Investors pushed Landed to double down on their brokerage service, which companies like Compass and Redfin were proving had venture-scale potential. "If that works and it brings in the kind of return we expect, that's what you guys should go do," Lofton remembers being told, even though he didn't have any experience running a brokerage business without a down-payment program attached to it. There was also the question of what would happen to Landed's thousands of existing customers if the company decided to pivot the business, and Lofton and his team were forced to leave. It may have meant their customers would be left in the lurch.

Ultimately, Lofton and his cofounder found a way to buy out the company's investors and retake control of the business themselves. It required Landed to lay off their entire staff, which had grown to about a hundred people, and Lofton is now servicing Landed's customers' down-payment products himself. He hasn't ruled out the possibility that he could raise more venture capital at some point down the road and give the big-scale idea a try when market conditions change. But for now, Lofton is happy with what the company has become, albeit disappointed he wasn't able to return capital to his original investment partners or to the countless staff and advisers who helped build the company along the way. He doesn't have any regrets about taking venture capital although he does see exactly how the incentive structure around venture-backed businesses could have forced them to make decisions that would have been worse for the homebuyers. "There were knobs that we hadn't dialed up as much as we could that we could see at some point they were going to ask us to turn," he said. One of them was on fees. "[Our investors] would say things like, 'As you get more traction and ability to push your prices, that's one area you're going to need to look at.'"

The lack of other options for capital that could have helped Landed weather a shift in market conditions is a problem that some investors are trying to fix. Michelle Boyd, the chief strategy officer at Terner Housing Labs, a nonprofit affiliate of UC Berkeley's Terner Center for Housing Innovation, is one of them. As a board member for Terner Labs, I am intimately familiar with the organization and its mission. Boyd, who has been advising Proptech startups with the aim of addressing the affordability crisis for almost a decade, started noticing a troubling pattern among those companies a few years ago. While the companies had solid business models that aligned profit with equitable outcomes for the housing market, they weren't able to find investors who thought the opportunity was big enough. As Boyd described the problem, "VC funders primarily support 'Prop-tech' software companies that bring efficiencies to the real estate market without fundamentally affecting affordability, equity, or sustainability outcomes. When VC investors do invest in more 'physical' innovations, they place immense pressure on these firms to scale rapidly, often at the cost of scaling well or sustaining a mission focus." For companies working to solve the housing crisis, one of the country's most intractable inequities and greatest threats to economic security, the impossible choice between venture funding that forces breakneck growth or no growth at all is particularly harsh.

Boyd's answer to this problem has been to band with other housing innovators to create a new investment vehicle, called Joist, which aligns investor outcomes with the mission-driven purpose of the companies. Joist began raising its first fund in 2024 and offers a unique mix of equity and debt that provides innovative housing startups, unable to raise venture capital or qualify for bank loans, the capital they need to bring their solutions to market on the timeline and at the scale that fits their purpose. By doing so, Boyd hopes Joist can show other investors that there are significant returns to be made by investing in companies that meaningfully address the housing affordability crisis.

While vehicles like Joist, which find ways to align solutions to the housing crisis with financial returns, are a promising development that can provide a template for others to follow, it remains to be seen whether these new methodologies can attract the support of institutional LPs. Joist's initial capital comes from philanthropic sources that are investing more because of the mission than the return. "It is the lowest-hanging and most risk tolerant capital to start with but it alone will not shift the market," Boyd says. "We believe there is an LP base out there. But I have seen other funds with a similar mission try to raise from those traditional LPs and end up falling into the typical Proptech trap. We want to make sure we have mission-aligned investors as we prove that these companies, with the right time and support, can build viable businesses that deliver a return and also solve a real problem." If Boyd is successful, Joist could shift the way housing construction is financed. But it will require LPs to shift their mindset, and as we will see in the next section, that won't be a simple task.

The future of Proptech is uncertain. Rising interest rates have had a more profound effect on Proptech than probably any other subsector of the tech industry. Venture-backed real estate companies have been shedding jobs at a faster rate than other corners of the tech sector and valuations are falling. Yet the fundamentals of the housing market remain such that there will be market opportunities to exploit for decades to come. While new housing construction has picked up slightly since the beginning of the pandemic, it is still far below pre–Great Recession levels. Housing prices have been remarkably resilient in the face of rising interest rates, signaling that there is an abundance of demand for housing that can't be met until supply catches up.

Perhaps the most sobering signal about where Proptech is going is Andreessen Horowitz's $350 million investment in Flow, a Proptech company started by WeWork founder Adam Neumann.

The investment, by all accounts Andreessen Horowitz's largest ever up to that point, was made in 2022 before the startup even had a product. Flow's exact business model remains opaque, though it appears to be something like a version of WeWork for the residential sector, combining some of the riskiest elements of fractional ownership, landlord tech, and shared equity all in one. Neumann has described Flow as a "consumer-facing residential brand" that is "integrating technology, community and a world class operating team that puts the resident first." Echoing Divvy cofounder Adena Hefets's pitch about co-ownership, he told the audience at an Andreessen Horowitz event in 2023, "if you're in your apartment building and you're a renter and your toilet gets clogged, you call the super. If you're in your own apartment, and you bought it and you own it, and your toilet gets clogged? You take the plunger. It's the difference from feeling like you own something to just feeling like you're renting." Neumann, who has taken very little responsibility for WeWork's collapse—and in fact recently attempted to repurchase the company from its current owners—has quietly been buying up thousands of apartment units across the South over the past few years. They are, presumably, going to be Flow's proving ground as the company rolls out its product suite.

Why Marc Andreessen decided to make such a big bet on one of the most notorious founders of the modern era seems baffling. But he made it clear when he invested in Flow that the way Neumann handled his time at WeWork was a feature, not a bug.

> *Adam is a visionary leader who revolutionized the second largest asset class in the world—commercial real estate—by bringing community and brand to an industry in which neither existed before. Adam, and the story of WeWork, have been exhaustively chronicled, analyzed,*

and fictionalized—sometimes accurately. For all the energy put into covering the story, it's often underappreciated that only one person has fundamentally redesigned the office experience and led a paradigm-changing global company in the process: Adam Neumann. We understand how difficult it is to build something like this and we love seeing repeat-founders build on past successes by growing from lessons learned. For Adam, the successes and lessons are plenty and we are excited to go on this journey with him and his colleagues building the future of living.

The housing market is the cornerstone of the American economy. It was responsible for creating the middle class and remains the most powerful symbol of the American Dream. At the same time, the commodification of housing in the US has exacerbated racial wealth inequality, produced the first generation of Americans who will, by many measures, be worse off financially than their parents, and pushed a record number of people into homelessness. Fifteen years after the foreclosure crisis, the new normal in housing is creating a threat to our economic prospects and driving a cynicism that is undermining our democracy. Venture-backed startups, by enabling Wall Street investment and toying around with new financial instruments, are destabilizing housing for everyday people. The returns that venture capital expects make the sheer size of the housing market an irresistible opportunity for investors, but their commitment to seek power law returns means the solutions they can fund are set up to exacerbate disparities in housing rather than reduce them. In the short term, some of these companies may not be able to survive high interest rates, wreaking havoc on the lives of their customers. In the long term, the persistent shortage of housing creates a systemic opportunity that investors will want to capitalize on. At the same time, companies that could provide much-needed innovation that may allow more people to access stable, affordable housing can't find the capital they need to scale their models. If we don't create guardrails around this new set of tools, constrain the ability of investors to exploit the market, and attract less extractive capital to the sector, venture capitalists alongside their private equity brethren could sow the seeds for the next great housing catastrophe.

5

THE PATHBREAKERS

CAN VENTURE CAPITAL
BE DONE DIFFERENTLY?

Housing is far from the only sector of the economy where the monoculture of venture capital drives negative outcomes. The unwavering commitment among venture capitalists—and limited partners—to pursue power law returns is distorting the entire innovation economy. Despite the hold that power law dogma has on Silicon Valley, the inconvenient truth is that *the power law approach doesn't actually work that well.* At least not at scale. When venture capitalists proclaim that, as an asset class, VC consistently outperforms the S&P 500 across the short and long term, what they aren't saying is that only a small handful of venture capital funds are driving that overperformance. The top twenty Silicon Valley firms realize 95 percent of all venture capital returns. In other words, power laws exist even across the ecosystem of venture capital funds, not just within them. When you take out the top 10 percent of all funds, the remaining 90 percent don't perform any better than the stock market. That is an astonishing imbalance that arises because there is a vanishingly small number of companies truly capable of achieving venture scale. Andreessen Horowitz operates on the assumption that just fifteen companies will be responsible for 97 percent of venture capital returns in any given year. Far less than 5 percent of venture-backed companies will become unicorns. Almost all the 1,000+ venture capital firms in the US are chasing that handful of deals. In turn, limited partners are doubling down on the firms that have proven they can land those deals, dumping more and more money into the top funds. This top heaviness makes the pressure to achieve outsized returns among those megafunds even greater. This results not just in the negative externalities, or the costs the rest of us are asked to bear, in areas like the housing and labor markets but also in a huge loss in economic potential among the companies that don't have access to startup capital, the ones whose founders have great business ideas and working business models—like LocalData, Good Eggs, and Foodnome—but can't find investors who will support their approach.

What if venture capital could be done differently? What if it were possible to deliver outsized returns investing in startups without the need to pursue power law returns? If we could, it would manifest a whole new asset class of companies that deliver real returns to investors—probably no worse than the average VC fund returns, and on a more consistent basis—without the need to grab every profit-maximizing opportunity no matter the cost. You could call it the Moneyball class of startups, winning with singles and doubles, in contrast to the bash ball that currently dominates VC.

The mainstream venture capitalists I spoke to were unambiguously dismissive of the idea. If it could be done, they say, someone would have proven it by now.

When I started writing this book, I assumed they must be right. There had to be robust and compelling data that proved the power law approach was the only way to win. Besides, I thought, venture capitalists are too smart—and greedy—to leave money on the table if it is there to be made.

But the more people I talked to, including the world's foremost experts on the history and practice of venture capital, the more it became clear that maybe the power law logic was flawed. While it might be true that, when all is said and done, the funds that make the most money have the most dramatic power law curves within them, that doesn't mean that the only way to achieve that success is to start with the distribution you want and try to reverse engineer it by shoehorning every company into one box.

For venture capital to work again, for it to surface and bring to market world-changing technology, we have to challenge this thinking.

WELCOME TO THE INDIE ERA OF STARTUPS

The irony of this groupthink within VC is that venture capitalists pride themselves on their outside-the-box thinking. The best venture capitalists, so it goes, see the potential in ideas that most people think are crazy. They're not afraid to go against the grain and challenge prosaic understandings of what is possible. When it comes to the approach they take to achieving returns, however, they are very much playing by the rule book.

Some startup founders are trying to challenge these orthodoxies. Jennifer Brandel is one of them. Brandel is, as she calls herself, an "accidental entrepreneur." She got the idea for her startup, Hearken, by trying to solve a problem she had herself. Working as a freelance journalist at a public radio station in Chicago in the early 2010s, Brandel was frustrated that the universe of voices she was able to hear from, and reflect in her reporting, was so small. She simply didn't have enough time or capacity to deeply listen to and reflect the points of view of a wide array of the communities where she was reporting. That lack of representation bothered her and she thought the emergence of new digital technologies, which were showing so much promise for lowering barriers for who was able to participate in the act of journalism, might be a tool she could use to augment her ability to hear from people and incorporate their perspectives into her stories. She got small grants from a few private foundations, one of which was also working with Corey Ford to get Matter.vc off the ground. Brandel wasn't thinking about Hearken as a startup at that point, but her program officer at that foundation convinced her to apply to Matter as a promising way to scale Hearken beyond her own newsroom.

She entered Matter skeptically. "I was really agnostic about what corporate form to take," she told me. "I just wanted to make something that's durable, that works, and that helps people." Recalling an interview during the application process, Brandel says she asked the selection panel: "It sounds like what you're looking for is me to convince you that I'm going to hit a home run, but what if all I want to hit is a double. Is that okay?" She recalls the room going silent and one of the interviewers saying, "Well, once you get in the program and see how it works maybe you'll change your ambition levels."

That made her uneasy, but despite her misgivings, she joined Matter in 2015. "I remember having cognitive dissonance but not knowing enough to challenge it," she says. Realizing that she didn't really know enough about how the startup and venture capital eco-system worked to critique it intelligently, she decided to stick with it and soak up as much as she could.

By 2016, though, she was getting frustrated. Not willing to shape-shift to fit the venture capital mold, she was having a hard time rais-ing money through the traditional routes. She also couldn't gain much traffic with social impact investors, who seemed more inter-ested in finding market-based solutions to problems in the developing world. It was at a gathering of social impact investors, her frustration levels at a peak, where she connected with Mara Zepeda, Aniyia Williams, and Astrid Scholz. The four of them were in the same pre-dicament: trying to grow solid businesses that solved real problems and created value in the world but finding themselves falling through the cracks—interested neither in venture-scale growth nor in oper-ating as a nonprofit.

As Scholz put it:

All of our companies were in a financing purgatory: Social impact in-vestors do not typically work in our spaces, and when they do, have narrow theories of change that do not contemplate the kind of system interventions we were making in our respective industries. And ven-ture capitalists won't fund companies that cannot or will not promise 10x returns, and that are run by people who do not look like them. In essence, we had created for-profit companies with purpose and then discovered this type of company is basically unfundable in the prevailing paradigm.

They knew there were more founders out there like them, so they decided to channel their frustration into doing what entrepreneurs do best: building a solution.

That solution became Zebras Unite, a community of like-minded founders, entrepreneurs, investors, and others in the startup ecosystem for whom the status quo wasn't working. Brandel and her cofounders picked "zebra" because it contrasted with "unicorns," the term used to describe the elusive billion-dollar-plus-valued startups that venture capitalists are chasing. "Unlike unicorns, zebras are real," they say. And, like a zebra's stripes, their companies represent two contrasting but interwoven concepts: profits *and* a commitment to positive social outcomes. In an alternative version of the future," they say, "we create a new model: We invent another style of investing, one that recognizes founders with long-term, wide-ranging visions. We focus not only on how much money is raised, and on investor returns, but also on how we generate value for users, elevate communities, and build a more equitable and inclusive system."

Their initial blog posts gained immediate traction. On their first two blog posts that outlined their thinking, they attached sign-up forms, thinking they would find a few dozen compatriots. The forms yielded thousands of submissions. Media organizations started picking it up. They received donations both from individuals and from foundations, including the one whose program officer had convinced Brandel to apply to Matter. The Zebras' first gathering, called Dazzle-Con (a group of zebras is called a dazzle), drew over a thousand people. The group has evolved into a worker-owned co-op and affiliated nonprofit, which creates a space for thousands of people representing dozens of chapters around the world. When I asked Brandel why she thinks the original blog posts hit such a nerve, she said, "I think it was tapping into a real spidey sense that so many mission-driven people had to be like 'Wait, there's no way for us to win, we're pawns in someone else's game.'"

Despite the growth of the Zebras community, they haven't spurred much increase in the flow of capital to Zebra-type companies. The funding environment for startups, especially tech-driven ones like Hearken, remains dominated by traditional venture capital and social impact investors who tend to treat their investing strategies as more akin to charity work. Without another option that can recognize and take advantage of the opportunity that all these doubles-hitting companies present—like LocalData, Good Eggs, and Foodnome—many of them will continue to struggle.

There are, however, a small but growing number of investors who are willing to challenge the power law doctrine. Their approach to venture capital may provide a blueprint for how to build a rich and diverse innovation ecosystem that overwhelmingly benefits society rather than exploiting it and delivers outsized returns in the process.

One of the most prolific of these investors is Bryce Roberts. Roberts, who runs a venture firm called Indie.vc, is hesitant to call himself a venture capitalist, but he has been in the startup investing game for longer than most other venture capitalists working today. He has a natural knack for entrepreneurship, which he began honing at the turn of the millennium at his first company, a custom ski equipment manufacturer, just after he left college in Utah. After leaving the ski company and doing a short stint at a startup that got purchased by Monster.com, he was asked by a friend who worked at a venture firm in Salt Lake City to take on a three-month project analyzing which of their portfolio companies were worth trying to save from the rubble of the dotcom bust. Roberts found that he was good at spotting trends, and the short-term gig turned into a full-time job at the firm.

In 2003, he and a friend organized a wildly popular tech business conference at which Tim O'Reilly, the venerated Silicon Valley publisher who ran a popular conference series of his own, spoke. O'Reilly was so impressed, he tried to buy the conference from Roberts and his friend. That didn't pan out, but it led to O'Reilly and Roberts starting a venture firm together, called O'Reilly Alpha Tech Ventures (OATV), in 2005.

Their timing was auspicious. As the infrastructure undergirding the internet was becoming more robust and the tools to develop and distribute software more accessible, the cost of starting a software company had fallen dramatically. This made it possible to invest much smaller amounts at much earlier exploratory phases to help developers figure out whether their ideas actually had the potential to become businesses. This was the birth of the seed stage, the earliest of venture stages, and OATV was one of the first firms to specialize in it.

As an investor in the youngest companies, Roberts was well positioned to observe startups' early growth trajectory. What he witnessed was troubling. With each subsequent funding round, he saw the entrepreneurs he'd believed in losing more and more control of their companies. Later-round investors weren't just taking more of the value of the companies, they were increasingly driving the decision-making from their seats on boards. He watched these founders face the same decisions that Alicia Rouault and Rob Spiro had to make, sacrificing what was best for the company—and often what was best for customers and society at large—at the altar of unfettered growth. He also recognized, much as Corey Ford at Matter was coming to realize at around the same time, that the overwhelming incentive for investors like him, who had gotten in at the earliest stages and were trying to raise more money from limited partners, was to push his portfolio companies to take on more and more risk in order to achieve the growth that would convince follow-on investors to place a higher valuation on the companies.

This dynamic throws founders into an inescapable cycle that requires them to keep raising money in order to become sustainable but see the goalposts get moved further and further away with every funding round.

That process, of continuously raising more venture capital in order to demonstrate value to future-round funders rather than focusing on building a solid business with strong fundamentals, is what creates bubbles. It is, more than any inherent risk associated with investing in startups, why Silicon Valley is such a boom-bust sector. Given what's at stake for venture capitalists, it is extremely difficult for founders to find off-ramps that might allow them to retain control of their companies and operate in accordance with what's best for customers, employees, and the long-term sustainability of the business instead of what will create the highest valuation in the venture capital marketplace. I have heard time and again from rueful founders, sounding a lot like recovering gambling addicts, how they thought every round of funding might finally be the one that allowed them to stand on their own two feet—only to be back on the pitch circuit again within eighteen months.

As interest rates began to rise in 2022, the folly of this approach was revealed. Companies whose unsustainable business models were being subsidized by cheap money from limited partners who were desperate to get a piece of the next runaway success were suddenly being asked to tighten their belts and focus on profit margins. Thousands of tech workers were laid off, even as the rest of the labor market saw record low unemployment rates. Many companies couldn't make the transition to a world where they had to stand on their own two feet. More than three thousand venture-backed startups went out of business in 2022, the highest failure rate since the Great Recession. While the pain was mostly confined to the tech industry, the excesses of Silicon Valley came dangerously close to taking down the banking sector when Silicon Valley Bank collapsed in early 2023.

It was this dynamic that made Roberts start thinking that there had to be a better way. What if founders had more flexibility to decide whether venture or some other form of capital was the right input to help them reach the next stage of their business? As Roberts put it, "We wanted founders to see beyond a binary world of 'going big or going home' into a world full of alternative outcomes that wouldn't require them to trade a lifetime of ownership and optionality for a seed round today." He convinced his partners at OATV to let him try something different with the new fund the firm was raising, its fourth. That fund became Indie.vc.

Roberts's decision to start Indie.vc, which launched in 2015, wasn't strictly a moral crusade. He was convinced that traditional venture capitalists were leaving a lot of money on the table. Roberts was watching the rise of the megafund and the consolidation of the venture industry behind a stricter and stricter adherence to pursuit of the power law, and he saw an opportunity to challenge some assumptions.

As he dug into the numbers, he saw the same thing that Evan

Armstrong saw when he looked at the Index Ventures fund that hit it big with their Figma investment but failed to return the fund. Roberts became even more convinced that the broader venture capital community was missing something big. He collaborated with Armstrong to publish an article in which they made their case:

> We have now reached a point in the startup ecosystem where for large VC funds, a startup achieving a **billion-dollar outcome is meaningless**. To hit a 3–5x return for a fund, a venture partnership is looking to partner with startups that can go public at north of $50B dollars. In the entire universe of public technology companies, there are only 48 public tech companies that are valued at over $50B. Simultaneously there are close to 1,000 venture funds all trying to find these select few ... To hit this $50B hurdle, entrepreneurs take on more and more risk to try and achieve larger and larger outcomes ... This is a market ripe for disruption. As venture funds continue to target larger and larger outcomes, there is a ton of opportunity left on the table that no one is seizing. You could invest in businesses that have an 80% chance of being worth $300M, rather than a 1% chance of being worth $80B. This strategy is an obvious opportunity to make a ton of money. Start by serving the underfunded, slowly move upmarket, and then, suddenly, you've disrupted the entire industry.

Roberts would target those companies, the 99.5 percent of startups that don't fit the venture capital mold, building his fund around those he thought could be doubles and triples rather than grand slams. Given that there are so many more of them than the ones that will return 100 times or more, he figured his odds were higher that he could make the same amount of money as a power-law-distributed fund without having to exert the kind of pressure to grow that founders often feel from their investors. If he did end up hitting a grand slam here or there, all the better, but he didn't need them.

He has likened this strategy to one that highly successful independent studios in Hollywood like A24, which made *Everything Everywhere All at Once*, take when they invest in lots of relatively low-budget movies that are ignored by the big studios, who need blockbuster outcomes to make their finances work. By using a discerning eye, these indie studios often end up knitting together portfolios of movies, cast aside by mainstream Hollywood, that do extremely well. Sometimes, as with *Everything Everywhere All at Once*, they end up hitting grand slams. But importantly, those grand slams aren't necessary for their strategy to make a lot of money for the studio overall. It's also notable that these smaller studios were able to meet the terms of the writers and actors unions while they were on strike in 2023, signaling that not only does their strategy result in business success but they can get there without exploiting workers.

Roberts thought this was an obviously compelling approach and that, once limited partners saw there were other options for private capital investment that could produce returns on par with traditional venture capital, they would flock. By 2022, Roberts says he had compelling data to prove he was on the right track. The two metrics by which venture capital funds are typically judged, internal rate of return (IRR) and "total value to paid in" (TVPI), were both higher for Indie than industry average. His internal rate of return, a metric used to approximate the annualized return rate of a venture fund, was 51 percent (anything over 30 percent is considered good in venture capital). He also claims the TVPI—the ratio of the fund's value to the amount limited partners had paid in—was over four to one, healthily within the 3–5x range investors shoot for. According to Roberts, 25 percent of his companies were bringing in revenue north of $5 million per year, and he even had a couple of "fund re-turners," those highly sought investments that, on their own, will deliver the entire value of what was paid into the fund.

One of his most successful investments is Nice Healthcare, founded by Thompson Aderinkomi, a serial entrepreneur who spent his early career working at a home health-care company. In 2011, his young son had a medical emergency that required a trip to a clinic where Aderinkomi had what he says was an "unpleasant, time-wasting and expensive experience." That trip to the doctor inspired him to start thinking about ways to fix the health-care system. Even-tually, he founded a company called RetraceHealth that offered an all-inclusive subscription model for access to a suite of health-care services like lab tests, imaging, and visits with nurse practitioners. Aderinkomi hoped to bootstrap the company, taking on personal debt to get the business off the ground. But it struggled initially and, with the high overhead costs of running a health-care business, Aderinkomi found himself needing to raise outside capital.

This was early 2016, when venture capital was at its peak popu-larity. Like many other young entrepreneurs at the time, Aderinkomi didn't think there were any other options for an early-stage tech company that needed startup capital. "The universe of information, people like you doing this kind of research, that just didn't exist in 2016," he told me when I talked to him recently. He needed only $500,000 to get to a stage where he thought the company could be self-sustaining, but he found it wasn't possible to raise such a small amount simply to get to sustainability. The venture capitalists he was pitching convinced him that he needed to push to be bigger, which meant taking on more money than he needed—in exchange for a bigger stake in the company. He wasn't thrilled about the idea, but he was running out of money fast and saw this as the only way to save his company. The move was successful, in its way; he raised $8 million over two rounds of funding and had convinced himself that once he hit profitability, he could leave venture capital behind for good.

It didn't take long for his vision for the company to come into conflict with what his investors were demanding. Blindsided just three months after closing his Series A round, his lead investor called to tell Aderinkomi that he and his cofounder were being fired. All Aderinkomi is able to say legally about the reasoning behind his firing is that it stemmed from a difference in perspective about a product decision. Aderinkomi wanted to do something he saw as truly solving a problem in the health-care space. His investors wanted him to take a path that, in his view, didn't break new ground but mimicked what other successful venture-backed health-care companies were doing.

With Aderinkomi and his cofounder out of the way, the investors were free to push the gas. But their attempt at Blitzscaling didn't pan out. Shortly after they fired Aderinkomi, RetraceHealth went out of business.

The experience left a bitter taste in Aderinkomi's mouth. Shortly after he was ousted, he came across Bryce and Indie.vc on Twitter. "Damn, I wish I would have found this guy six months ago!" he says he told himself. He took a break from entrepreneurship to spend time with his family, but it wasn't long before he got the entrepreneurial itch once again. He and two of his RetraceHealth employees started plotting a rebuild of the company they still wished they had had a chance to see through. He wanted to do it, but he was so disillusioned by his experience raising venture capital that he didn't think he could stomach the process of raising money. Until he remembered Indie.vc. He reached out to Roberts on Twitter five months after launching Nice Healthcare—a carbon copy of what Retrace-Health had been. Roberts was impressed by the early signs of success the company was showing and, after three half-hour phone calls, wrote Aderinkomi a check for $350,000.

Roberts's investing model requires a different set of tools from what traditional venture capitalists use. Typically, founders sign a term sheet that gives investors a percentage stake in the company that they cash in when there's an "exit"—the company either gets bought or goes public. Investors' stakes in their portfolio companies become more valuable as they raise future rounds of funding at higher valuations. Hence the reason earlier-stage investors push hard for companies to do the things that will make them look attractive to later-stage investors, and why it's extremely difficult for founders to get off the VC treadmill once they're on it. Roberts's thesis, on the other hand, was based on the idea that startups should have the agency they need to pursue whatever path is best for the long-term health of the business, not the short-term interests of investors. So if he was going to make money, he would need to find a way to align his return expectations with the business fundamentals of his portfolio companies.

His solution was to create an innovative new term sheet that includes a payback mechanism if the company decides to pursue sustainability rather than venture-scale growth. This term sheet looks a lot like the standard one that venture capitalists typically use, what's called a convertible note, which converts to regular equity if a company makes the decision to raise future rounds of venture capital. This is the standard way early-stage investment vehicles work. If the company decides not to raise more money, however, Roberts's term sheet gives founders the opportunity to take a different path. Instead of taking an equity stake in the company, Roberts would receive a cut of the company's revenue until he has earned a 3–5x multiple on his original investment. The return schedule is structured such that Roberts will earn his full return within eighteen to twenty-four months.

For Aderinkomi, this optionality was a game changer. Since Roberts did well whether Nice Healthcare raised another round of funding or not, Aderinkomi had the freedom to pursue the best outcomes for the business without the pressure to chase venture-scale growth. That freedom has led to a vastly different outcome than the one realized by RetraceHealth. Nice Healthcare has thrived, expanding operations to twelve states. In the wake of this success and armed with the leverage he would need to keep control of the company, he decided that the right path was to raise venture capital. In 2022, he raised a $30 million Series A funding round from a set of traditional venture capitalists, which allowed Nice Healthcare to expand around the country. Roberts's stake in Nice Healthcare now looks like any other early-stage investor's would in a traditional VC model.

Even though Roberts is no longer formally involved in Nice Healthcare's management (he served on the board until the Series A), the influence of Indie.vc's model is still being felt. Aderinkomi recently found himself in a disagreement with one of his investors about the path the company should take—a product decision, similar to what RetraceHealth faced before he was pushed out, that he felt would be bad for the business and customers—but this time he got his way. The control he maintained over company decision-making, because of the foundation Indie.vc had allowed him to build, meant he was able to choose a path defined by things other than what would be best for investors.

It is hard to understand Nice Healthcare's story as anything other than a win for everyone involved. The company is thriving. Roberts has the outlier success—the fund returner—that every VC craves. His limited partners are going to realize a venture-scale return on their investment. And the rest of us—especially the residents of Minnesota, where Nice Healthcare is based—get a successful, Black-led company that is creating jobs, adding to the tax base, and providing much-needed innovation in the health-care space.

Aderinkomi, for his part, has become an evangelist for Indie's model. "I wish every founder would get the chance to take money from Indie.vc," he has said.

Inexplicably, however, limited partners have been lukewarm at best on the Indie.vc approach. Despite his early success, limited partners started dropping out when they learned Roberts wasn't going to be chasing the next Uber. "We are out," one of these limited partners wrote to Roberts as he was raising his second Indie.vc fund. "The shift in strategy for the fund over time (for your good, intentional reasons) has moved further away from the kinds of companies we are looking to have exposure to." Unable to raise money from limited partners for his next fund, Roberts stopped investing for a while, and contemplated closing up shop altogether. Thankfully for him, for a new cohort of entrepreneurs, and for the communities that will benefit from the growth their companies will create, Roberts hung out his shingle again in 2023 and began making new investments in 2024.

Roberts's term sheet may be unique, but the general principle that drives his approach, something called revenue-based financing, has been gaining traction in the startup space. It's the idea that investors can make money by sharing in the revenue the company generates rather than simply its future growth.

Revenue-based financing isn't the only alternative to venture capital that has gained traction over the last decade. In 2012, as part of the Jumpstart Our Business Startups Act (JOBS Act), Congress loosened rules about who can invest in private companies. Prior to the JOBS Act, only "accredited investors"—wealthy people or institutions—could invest in private startups. It was seen as too risky for average people, though there was a vocal network of advocates (many of them in Silicon Valley) who thought that giving everyday investors access to the same wealth-building opportunities that rich people benefited from was the democratic thing to do (that democratization argument again). The idea didn't gain much traction in Washington until after the Great Recession when, seeking to stimulate small business growth, a bipartisan group of lawmakers championed the JOBS Act. The hesitancy to allow the masses to invest in private startups was understandable, but loosening investing rules gave entrepreneurs another option to raise money to get their business idea off the ground without having to seek traditional venture capital.

Many other entrepreneurs forgo investment altogether and decide instead to "bootstrap" their businesses, relying on the personal funds of the founders to get started and then using funds that the business has earned from customers to grow over time. Bootstrappers often count on family members, called a "friends and family round," to help fill in the gaps.

"We have an opportunity to change the meat industry and I can only say that because we don't have investors."
—MIKE SALGUERO, FOUNDER AND CEO, BUTCHERBOX

Some companies combine the concepts of bootstrapping and crowdfunding, using platforms like Kickstarter to presell their product. One of the most successful of these companies is ButcherBox, the subscription meat company. ButcherBox's founder, Mike Salguero, turned to Kickstarter after a bad experience with venture capital at his first company, CustomMade, in the early 2010s. Salguero relayed this story, and how it led him to forgo venture capital when he started ButcherBox, on the *How I Built This* podcast in 2022. According to Salguero, he and his cofounder, Seth Rosen, set out to raise venture capital right after the housing bubble burst in 2008. He had just been laid off from a real estate brokerage when he got an opportunity to purchase a woodworking listing service website named CustomMade.com. He was able to raise about $500,000 to take over the business but very quickly realized he would need much more than that in order to scale it. He and Rosen went hat in hand to dozens of investors who all told him no, until in one meeting he described the business not as a listing service—basically a glorified yellow pages—but as a marketplace that connected artisans with buyers. Instead of collecting listing fees from the woodworkers, as they had been doing, the investors perked up at the idea of their turning the company into a platform that took a cut of every transaction— that tried-and-true venture business model that investors find so lucrative. Salguero and Rosen raised almost $2 million from prestigious firms First Round Capital and Google Ventures to turn CustomMade into a marketplace.

Things did not go to plan for Salguero and Rosen. For marketplaces to work, the sellers on the platform have to be committed to building a business, and the artisans who used CustomMade were much more focused on building their crafts. It soon became apparent that the platform would never reach high enough transaction volume to scale the company. At one point CustomMade was spending $600,000 per month and bringing in less than $50,000. Salguero wanted to pivot the business back to what had been working—the listing service—but investors forced him to stick with the marketplace model. In their view, the problem wasn't with the business model, it was with the team behind it. Besides, if they pivoted away from the marketplace model, it would make it harder to sell the company or raise an additional round of funding at a higher valuation down the road.

Salguero says investors persuaded him that he needed to fire much of his staff—the team who had placed a lot of faith in Salguero and Rosen when they were just getting started—and replace them with hired guns with more experience in Silicon Valley. When the staffing changes didn't fix the problems in the business model, ultimately the investors turned on Salguero. He was pushed out as CEO just after his first child was born.

This experience left Salguero bitter about venture capital. He'd had to betray many of the people who helped him get to where he was, and he felt the business wasn't really serving the artisans who had relied on it. But he remained an entrepreneur at heart. Twenty-four hours after leaving CustomMade, he had already sniffed out another business opportunity.

Salguero and his wife, who has an autoimmune disease, were pursuing an anti-inflammatory diet to keep her condition in check. One hallmark of an anti-inflammatory diet is grass-fed beef, but Salguero and his wife found it expensive and inconvenient to procure. They would often have to purchase a half or a quarter of an entire cow rather than just a few cuts, and they were constantly trying to off-load meat to their friends and family. "It would be so much easier to get this delivered to my house," one of Salguero's friends told him as he was dropping cuts of meat at his friend's office. The light bulb went off and ButcherBox was born.

As he realized that ButcherBox could really be a business, Salguero was resolute that he would not raise venture capital again. Instead, he decided to try and prove he could build a successful, venture-scale business without the need to sacrifice control—and his ethics—in the process. This is when he turned to Kickstarter, a popular crowdfunding platform. Initially, his goal for the Kickstarter campaign was to raise $25,000 through presales, but by the end of the thirty-day campaign, he'd made $210,000 from over a thousand commitments, making $40,000 in profit in the process. Confident that he was onto something, Salguero used this small nest egg to get ButcherBox off the ground with no need to raise money from outside investors.

Despite not being venture-backed, ButcherBox went on to grow like a venture-scale startup. It has turned that $40,000 profit in 2015 into a company that was worth over $600 million by 2023 and still growing, a home run by anyone's estimation. Not only was Salguero able to achieve that success without taking venture capital, he believes that this outcome wouldn't have been possible if he had. Salguero watched other meal delivery services tank in value right as he was wrapping up the Kickstarter campaign, drying up the market for funding for direct-to-consumer food delivery companies. If he had been on the venture capital treadmill, reliant on being able to raise another round of venture funding to keep the business running at that time, he is certain he wouldn't have been able to do it and the company likely would have had to shut down. "I don't have investors . . . who are saying that we need to do things differently," he told *Harvard Business Review* in 2023. "And that enables us to give the mission some breathing room."

After a post-pandemic slowdown in 2023, when customers returned to shopping at grocery stores with more frequency, Salguero was reflective. While the company is still profitable, he knows that if he had investors on board, there would be much more urgency to cut costs and increase growth. Instead, he is able to invest in things that will be better for the long-term health of the company and its customers. "The difference between us [and venture-backed companies in the food business] is that we're not on a specific timeline," Salguero told me. "It totally frees you from the urgency and forces you to think about long-term sustainability."

One area that Salguero is focusing on with all that space and time is worker welfare in meatpacking plants, one of the most dangerous places to work in the world. ButcherBox is starting one of the first programs to improve safety in meatpacking and, Salguero says, "If it takes three years to set up, that's ok. It's good for business and it's good for the industry we're trying to disrupt."

He also doesn't believe that venture capital would have allowed him to stay true to his mission, to provide high-quality meat to consumers across the country, and to do so more responsibly. "We believe that we have a bigger mission than just driving as much profit as possible to our shareholders," he told me. "In our space, oftentimes the animal, the farmer, the environment, and the workers suffer in pursuit of profits. That's something I believe can change, and I can only say that because we don't have shareholders," he said. "When times are tough they cheapen the meat. The only way to cheapen the meat is either one, build efficiencies, which is not where people usually go to, or two, you cut corners, and you start to greenwash things and it's not really what you're saying it is. It's like a cheaper version of what you're saying it is. We will never take that step."

The way he talked about the program to help meatpacking workers called to mind the many conversations I'd had with executives at venture-backed companies over the years, ones that inspired me to write this book. I have struggled to understand why they couldn't find the time or space to prioritize external projects in the community that seemed on their face to be, as Salguero put it, good for business and for the industry they're trying to disrupt—not to mention good for society at large. This conversation with Salguero helped me to see what I couldn't see before: that they couldn't consider these opportunities because the internal forces driven by investor imperatives demanded that they didn't. Those forces required them to focus on maximizing short-term growth at the expense of everything else, even when those things were good for business.

For his part, Seth Rosen, Salguero's cofounder at CustomMade, managed to revive CustomMade and turn it into a successful business. But it required buying out all their traditional venture capital investors. Now, instead of acting as a marketplace that connects creators with customers, CustomMade does all the design and creation in house. Unlike marketplace businesses like Shef and Uber, CustomMade employs the artisans directly, an approach that is squarely at odds with the growth expectations of venture capitalists but that makes the most sense for the business. Rosen won't give too many details about how well the company is doing, other than that they are bringing in "well into eight figures of revenue" per year, a significant improvement from where they were in 2014, on the brink of shutting down. Rosen doesn't seem to harbor any ill will toward venture capitalists, but he also recognizes that it can be very difficult for entrepreneurs to discern when their business should seek venture and when another form of capital is the better choice. Echoing what I heard from Rob Spiro, "You have to match the right tools to the right job."

Each of the models outlined in this chapter shows promise for offering a path forward for entrepreneurs who don't want to give up control of their company or for whom the hyperscale model isn't appealing or doesn't fit their business. But these non-dilutive paths are significantly more difficult to traverse than the path that traditional venture capital offers. The JOBS Act places a cap on how much money a startup can raise via crowdfunding. Bootstrapping usually requires an entrepreneur to have the resources and wherewithal to personally underwrite the business. And then there's the fact that many of these companies have competitors who are taking huge sums from venture capitalists, threatening to undercut the business before it can gain a foothold. The founders of Tuft and Needle, the bootstrapped mattress startup, tell of a venture capitalist who came calling after the company made its first million dollars. The VC told them that another mattress company, which had raised millions from venture capitalists, was launching soon. "You guys should take the money too, or else you'll be squashed," he told them. They didn't take the money, and ended up selling the company to Serta Simmons in 2017, illustrating both that venture capital money is used to influence startups even when they aren't venture funded and that it is possible to build a successful startup without raising money from traditional VCs.

The platform that Salguero relied on to get ButcherBox off the ground—Kickstarter—is itself a successful example of how to build a venture-scale tech company without losing control to venture capitalists. Kickstarter was founded in 2009 by Perry Chen, Yancey Strickler, and Charles Adler. As an artist living in New Orleans in 2001, Chen got the idea for Kickstarter after he tried to convince a New York DJ to come play a gig at the New Orleans Jazz Fest. Chen couldn't make it work because he couldn't get the money together and he thought, what if there was a way for people to make a concrete pledge to buy a ticket to a show and if enough pledges were made, the show would move forward? He ultimately moved back to New York and connected with Strickler, a music writer, to flesh out the idea. Adler, a graphic designer, later joined the team to design the original version of the website.

As Strickler told me recently, the three thought of themselves as creatives, not as businesspeople. They were dismissive from the start of the profit motives of potential investors and were deeply committed to maintaining Kickstarter's mission of being a way to support creative projects that might otherwise never come to fruition. They pledged early on never to sell the company or take it public, severely limiting their upside potential. But they would need startup capital from somewhere if Kickstarter was to become more than just an idea. Initially, they relied on friends and family who chipped in a few tens of thousands of dollars. But to build the experience they knew they wanted to create, they ultimately realized they would need to expand their circle of investors.

They'd stacked the deck against themselves by removing the option to go public or get acquired by another company. They got lucky when they landed a meeting with Fred Wilson at Union Square Ventures, an iconoclastic venture capitalist who earned the respect of Silicon Valley with successful investments in brand-name companies like Twitter but who also has a knack for investing in companies, like

Etsy and Tumblr, that were founded by socially conscious entrepreneurs and that had a reputation for more responsible business practices. Wilson led Kickstarter's Series A funding round, which came in at less than a million dollars, a wildly small amount of money for a funding round even in 2009, when the economy was still reeling from the Great Recession. The effect of raising such a small amount of money—something that you'll recall Thompson Aderinkomi initially tried and failed to do—is that it allowed the Kickstarter founders to retain more control of the company's destiny. Wilson had a seat on the board, but the rest of the investors, who were all wealthy individuals rather than firms, didn't have enough skin in the game to justify butting in too much.

If you were on the internet in the 2010s, you know what happened next. Kickstarter pioneered the crowdfunding trend, one of the most important innovations of the social web era. It was soon funneling millions of dollars from small-dollar backers to thousands of creative projects and entrepreneurial ideas around the world. Some projects, like ButcherBox, allowed backers to place preorders. Other projects gave backers fun tokens in exchange for what was essentially a small-dollar donation to fund a cool idea. In all cases, it gave people a sense of buy-in and engagement with the act of cultural production and entrepreneurship—in many ways the true democratization that the internet had promised.

Chen, Adler, and Strickler soon became the darlings of the startup universe. But with great success came a lot of expectation that Kickstarter would follow the same path that its peers were taking. In 2012, Kickstarter's main competitor, Indiegogo, raised $15 million, followed eighteen months later by a $40 million Series B funding round. This gave them far more capital to work with than Kickstarter had, causing Strickler to question whether they needed to do the same thing. "I remember feeling like 'oh my god, we have to match that,'" he said. The pressure, according to Strickler, wasn't coming from investors but from employees who themselves were heavily influenced by the get-rich-quick narrative spun up around Silicon Valley startups. "Every day there was a story about how much someone raised," Strickler said. "Media more than anything influenced how people thought about these things. They made people think fundraising was success."

Because of their high-profile success (by 2012, Kickstarter was moving $1 million per day to its projects), Strickler and his cofounders had more freedom over their destiny than many startups of that era, and they wanted to make sure their commitment to the mission would last even beyond their time at the company. So they started thinking about how they could set up Kickstarter to be resilient against any outside forces or future leadership that wanted to prioritize profit over mission. They found their answer in a new form of corporate governance that was just being birthed.

In 2007, a group of friends with Wall Street backgrounds decided they wanted to use their finance skills for good. They thought there were too many companies prioritizing shareholder profits above all else and they wanted to create another way, a model that would allow executives to consider the effects of their businesses on other stakeholders—and society at large—in addition to the profits they were making for shareholders. This idea became a corporate certification program known as B Corp, which stands for benefit corporation, that assessed companies' commitment to social outcomes beyond the bottom line. The program gave high-scoring companies a badge they could use to market their world-positive values. The organization that emerged to run the certification program, B Lab, certified eighty-two companies in its first year.

The B Corp concept earned a lot of fair criticism early on for not having any teeth. But beyond the sense that B Corp was simply a superficial marketing ploy, another more fundamental challenge emerged. The general understanding of corporate law, defined in statutes as well as case law in all fifty states, is that companies have an obligation to make as much money as possible for their shareholders. To include anything else in their calculus could potentially open up B Corp companies to lawsuits from investors who thought their beneficence was leaving money on the table. For companies to feel safe that they could actually practice the tenets B Lab was proposing, they would need a separate governance form that made explicit their fiduciary responsibility to social outcomes as well as profit. The team at B Lab started approaching states to see if they would enshrine a new corporate governance form, the Public Benefit Corporation, or PBC, that allowed companies to define their stakeholders more broadly than just shareholders or investors.

The first state to take them up on it was Maryland in 2010. But few companies took advantage of the model until Delaware, which because of its low-overhead corporate legal structures is the country's most popular place to register a business, adopted it. Now, well over half of US states allow companies to adopt some form of the PBC model.

Kickstarter reincorporated as Kickstarter PBC in 2015. Its new charter outlines a set of values that guide the company's decision-making, including a recommitment to its original mission to support creative projects but also broader good governance principles like a commitment to avoid tax loopholes that would lower its tax payments and to limit its environmental impact.

For Strickler, who left Kickstarter in 2017 and is now embarking on another startup journey, this is a happy ending. "I remember see-

ing Blitzscaling and thinking 'we want to be the opposite of that,'"
he said. "Our goal is to be a small business. We want to be like the
corner bodega: You make more money than you spend, you're reliable, and people trust you. That's what equals a good business, not
that we crush our competition and own markets." They may not
have vanquished their competition, but Kickstarter is still doing remarkably well. Since its launch, almost 23 million people have sent
$7.5 billion to over 250,000 projects. The amount of money pledged
to Kickstarter projects has increased every year since it became a
PBC.

While PBCs are a positive development in corporate governance,
moving away from the misguided concept of shareholder supremacy that has dominated capitalism for the last century, they still
have significant shortcomings. The biggest is that they don't *require*
companies to behave a certain way. They just provide protection for
those executives who *choose* to put mission over profit. The companies that want to enact stricter protocols that mandate certain behavior no matter who is in charge are mostly left to create their own
governance structures.

There are a few notable examples of these bespoke structures in
tech, most famously the one employed by OpenAI, which puts the
for-profit entity that develops and markets ChatGPT under the control of a nonprofit whose mission is to "ensure that artificial general
intelligence benefits all of humanity." The company also places a cap
on the amount of returns that investors in the for-profit entity can
make, an interesting indicator that it understands just how much
investor returns can influence product and business model decisions.

OpenAI's model faced a constitutional crisis in late 2023 when the nonprofit's board fired its CEO, Sam Altman, sending the tech and business worlds into a tizzy just before Thanksgiving. Altman was ultimately rehired and most of the board resigned, but as of this writing, the governance structure that (in theory) protects the mission of the company remains in place.

I spoke with Altman a few weeks before the firing-rehiring drama played out, to try and understand why he—a prolific Silicon Valley investor himself—felt it necessary to create such a complicated structure that limited the involvement of investors. I was expecting him to be cagey and defensive about this, maybe to claim that the governance structure wasn't just about constraining the profit motives of investors but is also about reining in the capitalist motivation of everyone involved in the project. But when I talked to him, he was very forthcoming about his belief that VC pushes certain types of companies to create risks for society. "I am a huge believer in the idea that incentives are superpowers," he told me. "As Charlie Munger said, 'show me the incentive, and I'll show you the outcome.'"

One of OpenAI's main competitors, Anthropic AI (which was founded by a breakaway faction of OpenAI employees who were even more concerned about AI safety risks), also has constructed a bespoke governance model with the intention of protecting the company's mission from the vagaries of investor demands. Anthropic's model is a hybrid. They are incorporated as a Public Benefit Corporation in Delaware, but they have also created what they call a Long-Term Benefit Trust (LTBT) that, by 2027, will have the authority to select a majority of the company's board members. The trustees who oversee the LTBT are selected based on their commitment to and expertise around the safe deployment of artificial intelligence and will have no financial stake in the company. The terms of the trust arrangement also require the company to report to the trustees "actions that could significantly alter the corporation or its business." In devising this model, the company was trying to, as it announced in 2019, "align our corporate governance with our mission of developing and maintaining advanced AI for the long-term benefit of humanity."

Anthropic's reliance on a trust to mandate an adherence to its mission is novel but it isn't unique. Patagonia, for example, made headlines in 2022 for transferring all of its voting shares to a trust whose mission is to protect the environment and fight climate change. Big-name European companies like Novo Nordisk, Bosch, and Carlsberg all operate under similar structures. These models, where the controlling stock of a company is placed with an entity that is obligated to serve a mission rather than a set of owners, have gained traction in recent years in part because of the work done by a German organization named Purpose Economy, which advances what it calls "steward-ownership" corporate governance models.

The steward-ownership model that Purpose Economy advocates for is represented in the experience of one of its leaders, Maike Kauffmann. Kauffmann's mother owned a publishing business that advocated for organic food systems in the 1990s. When her business partner decided he was ready to retire, he knew he didn't want to sell his shares of the business to someone who might not appreciate the company's values. He wanted to ensure that the company's mission would remain intact no matter who owned it. There didn't seem to be an off-the-shelf model that he could employ, so he and Kauffmann's mother built one themselves. Its basic structure reflected the two main principles that define steward-ownership models. First, steward-ownership models place control of the company into the hands of people who are close to the business, not in the hands of external shareholders or absentee owners. This closely held control ensures that management and strategy decisions are made based on what is best for the business rather than what will maximize shareholder returns. Second, all profits generated by the company are reinvested in the business—capital investments of the kind that grow the real economy—rather than extracted by shareholders, fueling inequality.

This doesn't mean that steward-owned companies can't create significant returns for investors. It's just that those returns are what's called structured returns, that is, the upside return will be predetermined as part of the initial investment agreement. OpenAI's "capped-profit" model is a version of a structured return. Indie.vc's Bryce Roberts also builds a structured return option into his term sheet. These models allow investors to make a healthy return and for companies to remain focused on their mission at the same time, all the while creating real economic growth that benefits everyone.

As promising as these models are, what the OpenAI debacle demonstrates is that when investor incentives are not aligned with the mission that these bespoke structures seek to protect, that tension can easily undermine the mission. This tension only becomes stronger the more potential money there is to be made. It is one thing for Patagonia, an apparel company that has always been solely owned by the family of its founder, to be governed by a trust. But for extremely high growth companies like OpenAI and Anthropic, which are building breakthrough technologies that will reshape the global economy and have taken on billions of dollars from venture capitalists, that tension was bound to come to a head sooner rather than later.

The ability to raise money while adopting an alternative structure also reflects an enormous amount of privilege on the part of these companies' founders. The vast majority of entrepreneurs are not able to drive the kind of bargain Altman and the Anthropic team did with their investors, even in times when VCs have more money to invest than they know what to do with. Even Altman found it difficult, telling me, "It was very hard to raise under this structure. Most investors looked at it and said 'absolutely not, I'm not capping my profits.'" Creating a system in which any founder can do what Altman and his cofounders did will require much deeper structural change.

Yes, we need to make it easier for founders and entrepreneurs to pull these governance models off the shelf and deploy them without needing an army of lawyers to set up. But ultimately, the only sustainable solution—at least in the tech space—is to find ways to align investor incentives with the purported mission of companies like OpenAI and Anthropic. That means the ecosystem desperately needs more investors like Bryce Roberts at all stages of growth, but as we'll learn in the next chapter, it will also require governments to wield their authority, creating regulatory structures that produce a different set of incentives. Without these, no amount of corporate structure will insulate companies from the influence of venture capital.

6

THE CHANGEMAKERS

HOW TO SCALE WHAT WORKS

How do we get venture capital on track? How do we achieve the right mix of capital inputs that will create a healthy innovation ecosystem while also making sure that the appropriate guardrails are in place to prevent the negative externalities that are simply too prevalent among venture-backed startups today?

The answer, you might be surprised to hear me say, isn't to abolish venture capital as we know it. There are genuine grand slam business ideas out there, many of which come alongside breakthrough technologies that will vastly improve the human experience. Venture capitalists should be encouraged to try and find them. There are also many perfectly boring venture-backed software companies like Twilio and AppLovin that you probably haven't heard of but for whom the structure of venture capital was exactly the right fit to get them where they are—companies worth multibillions of dollars.

But for the far greater number of singles, doubles, and triples—the ones that venture capital is currently trying to reverse engineer into a winning lottery ticket—there are very few options that truly match their business potential. For innovation to thrive, we need venture capitalists to prioritize the pursuit of breakthroughs rather than the pursuit of windfalls. And we need to support other modes of investing that don't require strict adherence to power law returns in order to create value.

The good news is the solutions to these challenges are well within reach. While the problems I have described up to now are structural, as I mentioned at the beginning of this book, there are a handful of people who can begin to address what's broken about those structures simply by changing some of their own behavior.

Here I outline what those acts might be.

LIMITED PARTNERS MUST FIND THE COURAGE TO PURSUE OTHER PATHWAYS TO OUTSIZED RETURNS

"Reform to the system will only come from LPs."
—SAM ALTMAN

Far more than anyone else, limited partners hold the key to changing how venture capital operates. They are the linchpin of the system, deciding how money flows to which venture capital funds. They have the power to steer resources to different types of venture funds *and* influence mainstream venture capitalists to adopt new practices.

The good news is, we don't need to change anything in the regulatory environment to make it possible for limited partners to change their behavior. The bad news is that very few LPs are motivated to do so, even if they recognize the current state of venture capital as a problem. Some LPs are perfectly happy with the status quo and would like the power-law-pursuing approach to venture investing to extend even further. I don't think those LPs will be receptive to my arguments until there is a meaningful and sustained shift in the venture landscape, so let's leave them to the side for now.

This advice is for the mission-aligned LPs, the ones—such as pension funds, cultural institutions, universities, and private foundations—who exist (at least in theory) for purposes other than making wealthy people wealthier. They are currently sleeping on an opportunity not only to find new promising investment returns but to use their capital to spur the world-positive outcomes they claim to support.

There are a number of barriers to overcome within these institutions, which tend to be characterized by their conservative dispositions and slow-moving bureaucracies.

First, the people who are in positions of decision-making authority at these large institutions view themselves as caretakers of the pools of funds their institutions are investing, with a fiduciary responsibility to grow the corpus of these funds as much as possible to support the aims of the institution. It is not "their" money, so the mindset for these decision-makers leans very hard away from anything that may be viewed as experimental or outside the norm. The old management adage about no one ever getting fired for picking IBM holds particularly true in the world of these institutional limited partners.

Second, the decision-makers at these institutions are often several degrees of separation away from the fund managers who are actually building the investment portfolios. At large private foundations and universities, for example, the investment offices are usually separate fiefdoms far away from the parts of the institutions that are charged with carrying out the missions of the organizations. The investment offices, which are typically given enormous autonomy and leeway, in turn often farm out much of the actual investment activity to networks of dozens or hundreds of fund managers. Many of them in turn are employing their own networks of fund managers, adding yet another layer of distance between the investment decisions and the original source of capital.

This chasm between the core functional areas of the LP institutions and their investment functions can seem even wider for the administrators, deans, museum directors, and foundation presidents who are charged with carrying out the mission of the LP institutions but who typically don't hold a deep fluency in finance and capital markets. They also tend to be the kind of people—drawn to the work they do because of the impact they want to make in the world—who don't want financial realities to sully the mission-driven work they are doing. Taken together with the fact that these institutions are usually set up to discourage interactions between the investment offices and the mission-driven parts of the organization, they mostly do not concern themselves with what the other is doing.

Pension funds, some of the largest limited partners in the world, experience all these institutional challenges at an even more concentrated level. For starters, there is understandably more scrutiny over how pension funds are invested compared to private institutional endowments. In addition to the fiduciary responsibilities laid out in ERISA, pension funds are governed by an array of state laws as well.

While meant to protect pension funds from unscrupulous fund managers, in practice they have the unintended consequence of even more deeply alienating the people whose money is being invested from the choices fund managers make. Pension funds are governed by a group of trustees, usually including representatives from the workers whose retirement savings are in the fund. They are often made to feel that the process is too complicated to understand, a phenomenon that Vonda Brunsting, a longtime labor organizer who has worked with pension funds and is now director of the Global Workers' Capital Project at Harvard's Center for Labor and a Just Economy, told me one pension fund trustee called "intellectual bullying." Recalling the meeting where the trustee used this phrase, Brunsting said, "there was a murmur around the room, as trustees recognized how they too had been treated by the fund's staff and service providers."

Furthermore, the investment managers use the complicated regulations and financial jargon to scare pension fund trustees into believing that any deviation from the fund managers' recommendations could open the fund to legal liability. Brunsting describes an intractable network of financial and legal professionals around pension funds that make it much harder for pension fund fiduciaries—or anyone else—to influence change in the investing strategy. In addition to pressure from the fund managers, Brunsting says, "It's the lawyers, it's the consultants, that all come together to give the sense that it's better not to rock the boat."

Some limited partners are also terrified of being locked out of the best-performing venture capital funds if they develop a reputation for asking too many questions or putting conditions on their capital commitments. They have good reason to be fearful; when several states passed pension reform laws requiring fee transparency from firms that invested in public pension funds, they were locked out of blue-chip venture capital funds.

A third reason for the intransigence on the part of institutional limited partners is their unwavering commitment to a principle known as modern portfolio theory. Virtually all institutional investors strictly adhere to MPT, which says that the best way to maintain capital growth in perpetuity is to diversify asset allocation in such a way that maximizes returns while optimizing risk. This requires investors to strike a balance within their overall portfolio of low-risk, lower-return assets with high-risk, higher-returning assets. Allocations to venture capital, which not only balance risk/return profiles but also tend to perform inversely to other asset types (i.e., when bonds are up, venture capital is down, and vice versa), are a critical component of an optimized modern portfolio.

Venture capital also happens to be quite a small proportion of most institutional investment portfolios. The average commitment to venture capital across higher education endowments, for example, is 7.7 percent, compared to over 40 percent to traditional public stocks. As a result of this relatively small yet critical role that venture capital plays, those on the limited partner side aren't motivated to dig too deeply into the details. You can think of venture capital as the salt in the overall investment portfolio recipe. The right (or wrong) allocation of salt in a dish will absolutely make or break its quality. But as a proportion of the overall food on the plate, it is insignificant. How much time are you willing to spend to understand the sourcing of your salt, in comparison to your meat or produce?

For all these reasons, in addition to the fact that institutional investors are huge bureaucracies that are designed to resist change, moving limited partners to action will be no small undertaking.

There is, however, one subset of limited partners who may be convinced to move more quickly and, in so doing, could chart a path for others to follow. These are the specific set of private foundations that run grantmaking programs that aim to create a more responsible tech sector. They grant tens of millions of dollars annually to hundreds of organizations that are working valiantly to steer the direction of technological development toward more positive outcomes. In 2015, a set of them, including the Ford Foundation, the John D. and Catherine T. MacArthur Foundation (both of whom have funded TechEquity), and George Soros's Open Society Foundations—three of the largest private foundations in the world with almost $50 billion in their collective endowments at the time of this writing—organized a collaboration called the NetGain Partnership. NetGain seeks "to advance the public interest in the digital age." Over the last decade it has worked on issues like disinformation, digital security, and the threat posed by generative AI. More recently, a group of ten private foundations, most of which are members of the NetGain Partnership, announced a high-profile partnership with President Biden's White House "to ensure that AI advances the public interest," pledging $200 million to support organizations who are working to prevent AI from harming democracy and the economy.

At the same time, however, while the grantmakers at these foundations continue to spend millions of dollars addressing the downstream effects of irresponsible tech companies, the investment managers on the other side of the organization are investing heavily in the venture capital funds that back the business models that cause the problems in the first place.

This problem of private foundations using their investment strategies to undercut their own missions has long been a frustration in nonprofit circles. It stems from the way private institutional philanthropy came to be, arising out of the Gilded Age when robber barons like John D. Rockefeller sought to turn their massive fortunes into vehicles for philanthropy. They lobbied hard, at a time when skepticism about the role of private philanthropy in American society was much higher than it is now, to establish a corporate form and associated section of the tax code that gave them enormous leeway to use their private fortunes to steer societal outcomes and simultaneously burnish their reputations through charity.

One major feature of the structure they managed to create is an exemption from most taxes for these charitable vehicles, on the condition that the foundations spend 5 percent of their net assets every year to advance their public-serving missions. That 5 percent requirement, which includes both the grants the foundations make to charities and the programmatic expenses associated with that grantmaking activity, has remained unchanged since the early twentieth century. Foundations have control over what happens with the remaining 95 percent, and most of them choose to keep investing it in assets that will ensure the original corpus continues to grow.

Over time, much as with university endowments and pension funds, those who are charged with maintaining foundation assets have often elevated their fiduciary responsibility above their responsibility to the mission of the organization. Over the past several decades the investment returns of these private foundation endowments have far surpassed the 5 percent they give away every year. The pioneering Rockefeller Foundation has an endowment that is now worth twice what it was when it was founded, in inflation-adjusted dollars. The Ford Foundation, which is second only to the Gates Foundation in the size of its wealth, has an endowment worth $16 billion as of this writing—three times as much as its original bequest in inflation-adjusted dollars.

Part of this is justifiable: the larger the endowment becomes, the further that 5 percent allocation will go. But some of it is a reflection of the IBM problem: no one wants to be the person who made a decision that went against the grain and have it not pay off. The reputations of university and foundation presidents and boards are made in large part based on how large the endowments grow under their tenure. There is very little incentive for them to rock the boat.

Despite the rigid hold that these outmoded conceptions of fiduciary duty have on decision-makers within private philanthropy—and at other mission-driven limited partner institutions—there are signs that things may be shifting. The Ford and MacArthur foundations have been leaders in what's called mission-related investing, carving out money from their endowments to invest more directly in funds and companies with a social change mission. Both foundations have also recently made commitments to divest their portfolios of carbon polluting assets and divert those funds into assets that promote clean energy and climate change solutions. This investment strategy aligns with the foundations' programmatic commitments to address climate change. "Foundations like MacArthur use massive tax-exempt wealth as the engine to provide grants for the public good," MacArthur president John Palfrey said at the time the reallocation strategy was announced. "How we deploy our wealth through grants and impact investments makes a difference. How we invest and grow our endowment is important, too."

Some foundations have taken things several steps further, by shifting their entire endowment into assets that align with the programmatic mission. One of these, the Wallace Global Fund, is a member of the NetGain Partnership and the White House's AI funder consortium alongside Ford and MacArthur. The Wallace Global Fund, whose namesake, Henry A. Wallace, was FDR's vice president and an advocate for racial and social justice, made a commitment in 2010 to reinvest the entirety of its more than $125 million endowment into vehicles that aligned with its mission and grantmaking priorities. This marks a notable difference from other private foundations whose mission-related investments are typically more like 5 or 10 percent of the endowment corpus.

While the Wallace Global Fund's strategy shift mostly focused on divesting from fossil fuels and investing in clean energy, it demonstrated to its peers that it is possible to invest an endowment in a way that aligns with purpose without having to sacrifice investment returns in the process. According to data from the fund, their endowment has performed slightly better than a typical investment portfolio in both the short and long term and has seen returns that are as good as those of the top private foundations.

These commitments give me hope that change within these institutions is possible, and that there is a path forward not just to divest from specific asset areas like fossil fuels, but to invest in a different hypothesis for how we fund the innovation ecosystem. As a result, we can make significant progress to ensure that the tech industry's growth is aligned with the public interest. I am calling for these foundations to make a bet, a bet that there may be more than one type of way to fund innovation, and that those other ways may not just be more beneficial for society but might also result in better portfolio performance as well. As I've noted, the dirty secret of venture capital is that most VC funds don't perform better than the stock market. If a few of the largest institutional LPs moved their allocation from power-law-committed fund managers who are underperforming to those attempting something like the Indie.vc approach, they have very little to lose. At best, they have first-mover access to what amounts to a brand-new asset class—innovative start-ups with a slightly smaller market opportunity and slower growth trajectory than the stereotypical venture unicorn—and all the upside it may offer. For institutions that claim to be working in the public interest, most especially those that claim to want to build a more ethical tech sector, there is no excuse not to try.

There are promising signs that limited partners are starting to realize their power to steer venture investing toward world-positive outcomes. Though the concept of ESG, or environmental, social, and governance, investing has become politically controversial over the last several years, many limited partners and institutional investors remain committed to its basic principles. Adherents to these principles believe that by investing in companies and sectors that have more world-positive, sustainable, and responsible business models, they can make better returns over the long run. These beliefs have become surprisingly mainstream in the finance community. In a 2022 survey by PwC, 90 percent of asset managers believed investing in ESG funds would improve their overall returns, and the majority of institutional investors can point to data showing that it already has.

While ESG principles have mostly been defined (to the extent that they have been defined) in the context of public market investing, some researchers and advocates are recognizing the need to establish ESG principles that are appropriate for private markets, and venture capital in particular. These principles, like responsibility for product harms and ethical labor practices, cover just about every venture-driven harm I've discussed in this book. And yet, proponents of ESG principles for venture funds are not explicit about the fact that the demands placed on venture-backed companies to produce outsized returns make it very difficult for them to meet these ESG standards; the structure of traditional venture capital is fundamentally incompatible with ESG.

For limited partners who care about ESG investing, the way to address this incompatibility is to invest in funds that are organized around non–power law return profiles. Committing to funds that don't create hyperscale mandates for their portfolio companies are much more likely to result in positive ESG outcomes in all three areas that ESG covers: environmental, social, and governance. Companies that grow more thoughtfully and sustainably are likely to have a smaller climate impact. There are signs that an Indie-style approach to investing may improve diversity outcomes as well: 40 percent of the companies that Bryce Roberts invests in are led by people of color. These companies also are less likely to employ business models that require economic exploitation to work. And the governance structures these companies are free to adopt can create more accountability, limiting the possibility for fraud and other bad behavior by company leaders. Diversifying the startup investment ecosystem so that companies like this can access the capital they need to thrive—at all stages of their growth cycles—will be a critically important component of successful long-term ESG investing strategies.

While limited partners certainly have the most power to effect change in the VC ecosystem, there are other actors who also can help move the needle, including venture capitalists themselves, founders (and the people who work for them), and, of course, the government.

VENTURE CAPITALISTS SHOULD TEST DIFFERENT APPROACHES TO FUND CONSTRUCTION

There's a chicken-and-egg problem associated with catalyzing a new ecosystem of startup investors who follow the Indie approach. As one executive at a mission-aligned limited partner explained to me during my research for this book, even if he could get his institution to change its investing approach, there aren't nearly enough fund managers operating today who are investing according to Moneyball principles. While Bryce Roberts certainly isn't alone in his commitment to supporting base hits—other firms like Tiny Seed and Mozilla Ventures have taken a similar approach—his community can hardly be described as a healthy ecosystem. The funds that do exist tend to fund at the earliest stages of the startup life cycle, leaving few options for companies seeking equity investment as they grow.

What would it take to attract more fund managers to try the Indie approach? How might we seed the field so that limited partners have more options to deploy capital? The solution here may also lie with the private foundation limited partners to not only change their investment policies but to use their grantmaking function to seed the ecosystem. They can and should invest in organizations that are doing the work to build the movement.

YOUNG ENTREPRENEURS–AND THE PEOPLE WHO WORK FOR THEM–NEED DIFFERENT MODELS FOR WHAT SUCCESS LOOKS LIKE

One of the most frustrating aspects of writing this book has been a common refrain I hear from entrepreneurs who say, often after they have had a bad experience with mainstream venture capital, that they didn't know there was another path to take to achieve startup success. I have thought a lot over the last few years about that conversation I had with Alicia Rouault when she was making the decision to wind down LocalData. How could someone who built a product that clearly met a need, and had figured out how to create a profitable business around it, ever think they were a failure? Something is deeply wrong with an economy where people like her have internalized a notion of entrepreneurship that views anything other than multibillion-dollar outcomes as failure. Creating a market for alternative investment approaches will require not just an increase in the supply side, activating more investors to change their models. It will also require a shift on the demand side, from entrepreneurs who are intentionally seeking out the right kind of capital for their business.

For now, too many of these founders are unaware that these are questions they should even be asking, and that's not entirely the fault of venture capitalists. The tech press has long been willing to unquestioningly lionize the companies and founders who achieve huge valuations at light speed, ignoring the slow and steady trajectories of companies with lower valuations but more staying power. Some business schools are to blame as well, reinforcing the ideas behind the power law methodology in the classroom. Both types of institutions would serve these founders better by giving equal airtime to success stories that fall outside the typical model.

But even when entrepreneurs know about alternative fund options, they are hesitant to accept their terms, assuming it will make it harder for them to raise money from mainstream investors if and when they need to. This is a real fear and speaks to the need for a more robust alternative ecosystem to ensure founders that there will be aligned capital available to them at every stage of their growth.

One other group of people is also to blame for pushing founders to accept venture capital's terms for achieving success: the employees who work for startups. Venture-backed startups often attract talent with promises of potential riches down the line when the company hits it big. What the startups can't offer in salary will be made up for with equity in the company. This often puts employee shareholders in a position of agitating for the same kind of breakneck growth, damn the consequences, that investors demand. When OpenAI experienced its constitutional crisis in late 2023, it was the threat from almost all the company's employees that ultimately pushed the board to back down from their mission-protecting position and rehire Sam Altman. Some reporting at the time indicated that much of what was driving this overwhelming show of support for Altman over the board was that OpenAI's staff was worried about what would happen to the value of their shares if Altman was forced out. When I talked to Yancey Strickler about the pressures he faced to push Kickstarter to seek venture-scale growth, he told me that much of it came not from investors but from his employees who were eager to earn venture-scale payouts.

We aren't going to realistically change the dynamics of the tech labor force such that employees are happy to work for a normal salary. Not when other venture-backed companies continue to pay out in equity. But companies can experiment with other forms of compensation that align employee interests with the future growth potential of the startup. ButcherBox, for example, found a way to compensate early employees with equity, but then bought out employees at a very generous valuation using the company's profits—a form of revenue-based financing where the employees are "investing" with their labor rather than with capital from limited partners. As models like Bryce Roberts's Indie.vc gain traction, ideas for how to make sure employees are cut in on these deals should be part of the conversation.

Ironically, the recent wave of layoffs in tech may also change how employees view the hypergrowth path to value creation. Many of them accepted jobs at high-flying startups based on promises that valuations would continue to expand. When the venture tide went out, and the shaky foundations of many of those companies were exposed, those workers lost not only the value of their equity but, in many cases, their jobs. There may now be a stronger case to make that workers have a better chance at an equity payout at slower-growth companies, changing the dynamics of the talent marketplace.

GOVERNMENT CAN PROVIDE SOME CARROTS TO ENCOURAGE A HEALTHY RISK CAPITAL ECOSYSTEM—AND ENACT SENSIBLE GUARDRAILS THAT LIMIT THE WORST ABUSES

Government intervention isn't required for any of the actors discussed here to change their behavior, and to be honest, I'm hesitant to propose public policy solutions that try to prevent bad behavior through punitive measures. I'm not naive enough to think that the forces of capitalism won't find a way to prevail, regardless of how much I believe we should empower regulators and enforcement agencies. But I do believe there is a lot the government can do to incentivize good behavior by seeding the market for Indie-style investment approaches that, if successful, could cause the market to shift on its own.

At the same time, the massive returns venture capital creates aren't free. It's just that the cost is borne by the rest of us: potential homeowners, workers, small business owners. In other words, society at large. For capitalism to work, the people who are pushing the riskiest activities need to bear the brunt of the cost when it goes wrong, and it is government's role to ensure that accountability exists.

THE SMALL BUSINESS ADMINISTRATION CAN ALLOCATE CAPITAL TO FUNDS PURSUING ALTERNATIVE PORTFOLIO RETURNS

In the 1940s, solving the problem of the lack of capital for innovative startups had largely fallen to the private sector. Then, amidst the intensifying Cold War years of the 1950s, calls for the US government to invest in its innovation ecosystem became louder. It was around this time that NASA was founded in response to the Soviet launch of Sputnik. Major investments were underway in technology that would evolve into the internet as we know it. There was also a political and ideological imperative to invest in the private sector, to demonstrate the supremacy of capitalist systems over communist ones to serve as a foundation for vibrant societies.

In 1958, leaders in Congress passed the Small Business Investment Act to give the brand-new Small Business Administration the ability to inject capital into early-stage companies. The SBIA established the Small Business Investment Company (SBIC), whose role was that of a quasi-limited partner, making loans to investors who would in turn invest in startups.

The program was then and is still now somewhat controversial. Free marketeers weren't particularly enthusiastic about what they saw as a government incursion into business activities. In their minds, the program went beyond the appropriate role of government intervention in the market, preferring that government lower taxes and limit regulation instead. The program also suffered from some early hiccups as some fund managers exploited loopholes that enabled fraud, and others used the funds for speculative investments in areas like real estate instead of their intended purpose to support the startup ecosystem. And, naturally, this being a government-run program, there were issues with red tape and bureaucracy that turned a lot of fund managers away from the SBICs.

But, despite its flaws, the SBIC program may have done more than any other act of government, aside from relaxing the ERISA prudent man rule, to spur the modern venture capital industry. William Draper III, a venture capital pioneer who founded the iconic Silicon Valley venture capital firm Sutter Hill, credits the SBIC program with catalyzing his own successful investing career. He took $300,000 from the program, one of the first loans the SBIC ever made in the late 1950s, and turned it into almost $30 billion in investments over the next fifty years. High-profile Silicon Valley companies like Intel, Tesla, and Apple, as well as nontech companies like FedEx and Whole Foods, received investments from SBICs in their early days. Data from the National Association of Small Business Investment Companies suggests that SBIC funding led companies who received SBIC investments to create more economic growth, including higher job creation, than their peers.

Despite these bright spots, the SBIC program failed to achieve widespread appeal in the world of venture capital. Amidst a recession in the 1960s and, later, the ERISA rule changes that allowed pension funds and other institutional investors to participate in venture capital funds, most mainstream venture firms opted for private sources of investment in order to avoid the overhead associated with what was, despite its aspirations, still a government-run bureaucracy.

The SBIC program has also, up to very recently, had a major design flaw making it particularly unattractive to early-stage venture capital investors: the capital contributions were structured as loans, rather than equity, and required investors to make semiannual debt payments to service those loans. This was a big problem for investors whose portfolios were by nature highly illiquid. In a normal venture capital fund, limited partners aren't expecting a return for several years and there is no expectation that they will receive income from the funds until a company within the fund realizes a liquidity event.

The SBIC program, for its part, recognized that the rules made it unappealing for investors to access, and has recently moved to make adjustments that are better suited to the way venture capital investing works. In August 2023, the Small Business Administration created an "accrual debenture" that, instead of requiring semiannual payments, allows investors to wait to make those payments until a distribution event—a time when they are distributing gains to their limited partners.

According to the SBA, the rule changes are meant not only to make the SBIC program more aligned with the practice of startup investing; they also reflect a recognition that there is a big gap in the startup financing landscape to fund the kinds of companies hitting doubles and triples that are currently being left behind. As the

SBA put it in the announcement of the rule changes, these types of companies "are not sufficiently financed by private market investors due to lack of access, duration of investment, risk/return profile, or magnitude of capital required."

As I'm writing this, the new accrual debenture rules have just been rolled out. It's still too early to say what impact they might have on startup investment activity. The founders and investors I've talked to, even the ones who fit the profile of the kind of investor the SBA is trying to reach, mostly don't have the SBIC program on their radar and certainly haven't heard about this change to the structure of the program. But in the same way that the creation of the SBIC program in the 1950s played a huge role in spurring the growth of the venture capital sector, this new rule has the potential to spur similar growth of an investor class interested in funding startups that don't fit the typical venture-scale mold. Funds like Joist, the one Michelle Boyd has established to invest in companies addressing the housing affordability crisis, could find the capital they need to prove their approach works, attracting other LPs in the process. The SBA can and should invest in a much broader awareness campaign and seek to partner with LPs like the private foundations I mentioned above to help this capital take root where it has the potential to show that other models for startup investing can be just as successful as the power law approach.

THE GOVERNMENT CAN USE THE TAX CODE TO INCENTIVIZE AGAINST LARGE FUND SIZES

Aside from the unyielding pursuit of power law returns, the increasing size and the short time horizon of venture capital funds are the two biggest factors contributing to the harms that venture capital causes. The government could shift the incentives for both these issues by making tweaks to the tax code.

As I pointed out in chapter 1, the way venture capitalists are compensated—with a 2 percent annual fee regardless of fund performance, alongside a 20 percent stake in the fund's returns—is rife with incentives to push fund sizes up and operate on a very short timescale. Fixing the structural problems with venture capital will require addressing both of these issues, and there are ways to use the tax code to shift investor behavior.

There is nothing sacrosanct about the two-and-twenty fee structure. As Josh Lerner and Victoria Ivashina point out in *Patient Capital*, while fund managers have always charged a fee to cover the cost of their operations, the earliest funds pegged their fee to the actual cost of running their firms. Draper, Gaither & Anderson as well as Greylock, two of the first modern venture capital firms, produced budgets that they shared with their LPs, and charged fees that covered the costs outlined in the budget. As venture capital became more dominant and the balance of power between limited partners and venture capitalists tilted toward the VCs, this practice faded away and was replaced almost universally with the two-and-twenty model (or more, for the most successful firms). Greylock still charges their fee based on a negotiated budget, but they are the rare exception.

The problem of runaway fees—and the megafunds that make them possible—is unlikely to change organically. Yes, the venture capital downturn in 2023 created a small reduction in the target size of many new funds (though they remain in the multibillions of dollars). But limited partners aren't going to keep their money on the sidelines forever, especially as an AI-driven investing boom seems imminent. As long as the very strong incentive exists to make funds as large as possible, venture firms will continue to raise as much money as they can.

There are, however, ways to incentivize venture firms to raise smaller funds. Currently, venture capital and private equity fund managers benefit from a deeply inequitable quirk of the tax code called the carried interest loophole. Carried interest is the 20 percent stake that fund managers have in the returns on the funds they manage. Even though carried interest is essentially a performance bonus and not a capital asset, it is taxed at capital gains rates instead of income tax rates like the performance bonuses that most workers are paid. The argument for keeping capital gains tax rates low is that it encourages economic activity by giving people who own assets a reason to invest them. But fund managers don't own the assets they are investing, the limited partners do. On top of that, there is very little evidence that this kind of tax break incentivizes investment activity that wouldn't have happened otherwise, and there is plenty of evidence that these kinds of tax breaks have contributed to skyrocketing inequality.

Getting rid of the carried interest loophole would be the best solution to address the inequality it has created. But Wall Street and Silicon Valley have fought very hard to maintain it, beating back attempts to eliminate it at least three times in the past three decades. As long as it remains politically unfeasible to get rid of it, there may be a way to shape it to work at least partly to our advantage: make it available only to fund managers who raise funds below a certain threshold.

ADDRESS THE MORAL HAZARD CREATED
BY SHORT FUND CYCLES

Just as there is no science behind the venture capital fee structure, there's no good reason to set fund lengths at about ten years. I've documented the many abuses and negative externalities that result from the short-term mindset of venture-backed companies, and the cost that society bears when venture investors apply this short-term mindset in areas like housing and climate, which necessarily operate on longer timelines.

One way to encourage longer time horizons for venture funds—and to address the moral hazard that short-termism creates—is to force early-stage investors to hold their stakes in portfolio companies for longer periods of time. The current post-IPO lockup period is far too short, and there is no legal requirement that there be any lockup period at all (right now these lockup periods are enforced only through private contracts generated as part of the IPO). If early-stage investors are forced to consider the value of their shares at a time when the public markets have determined a company's true value—that is, when early investors are forced to care about what the market thinks of their company rather than simply what the next round of private investors think—they may be more likely to focus on creating sustainable businesses from the start. They may take into consideration things other than speed of growth and market share as they shepherd the company through its early days. This would give more founders at least a modicum of the freedom that founders like Mike Salguero at ButcherBox have, to think more expansively about what it means to create value in the world.

The short time horizon for funds also has implications for competition in the marketplace, and our ability to enforce antitrust laws.

Matt Wansley and Sam Weinstein, two law professors at Cardozo School of Law, published a paper in 2023 arguing that the venture capital approach to grabbing market share by subsidizing the true cost of a product in order to drive competition out of the market, as Uber and many other venture-backed marketplace companies have done, is a violation of antitrust laws. This cynical growth strategy, they say, is a form of predatory pricing, where companies artificially lower their prices to drive out competition, then jack them back up after they have secured a monopoly.

Venture-backed companies get away with this because the Supreme Court determined that in order for pricing to be considered predatory, a company engaging in the practice must be in a position to recoup the money they lost when they undercut price. There have been few, if any, successful predatory pricing suits brought since the Supreme Court issued that ruling in 1986.

But Wansley and Weinstein point out that there are actors who do recoup their losses, many times over, when venture-backed companies engage in predatory pricing: the venture capitalists themselves. They need only to pump the company up to be big enough that later-stage investors believe its market size (gained through artificial subsidization of the product's true cost) will translate into gains at some future point. The VCs then cash out before the public markets have a chance to determine the company's true value, and they don't have to bear the responsibility for the fact that the company will never recoup what it lost through the subsidies. "Will Uber ever recoup the losses from its sustained predation?" the authors ask. "We do not know. Our point is that, *from the perspective of the VCs who funded the predation, it does not matter.* All that matters is that investors were willing to buy the VCs' shares at a high price."

Lina Khan's Federal Trade Commission has been trying valiantly to apply antitrust laws to Big Tech in order to create some account-ability for the industry. Turning the antitrust lens from the compa-nies to their investors may be a more fruitful avenue to rein in tech's excesses and present one of the only viable pathways to hold venture capitalists accountable for the negative externalities they create. But it might also be possible to achieve the same outcome by extending the post-IPO lockup period so that these investors aren't in a position to recoup their losses by selling shares at inflated value.

SHINE A LIGHT ON THE VENTURE INDUSTRY, AND HOLD VENTURE CAPITALISTS ACCOUNTABLE FOR THE HARMS THEY ENABLE

Since regulators assume that sophisticated investors like limited partners are able to look after themselves, the venture capital indus-try, and private markets generally, are very lightly regulated. While investors must register with the SEC, there are almost no transpar-ency or reporting requirements, accountability mechanisms, or other guardrails around venture capital. But in the wake of the FTX im-plosion, there is increased interest from regulators to create more accountability for VC firms. As has been well documented, before FTX collapsed under a clownishly executed and not very well dis-guised fraud, it was the belle of the VC ball. It had attracted huge investments on very favorable terms from the most prestigious VC firms in Silicon Valley. Sequoia Capital, which invested $214 million in FTX as it celebrated Sam Bankman-Fried's genius, lost all of it when the company collapsed (though, ironically, it appears all of FTX's creditors will be made whole due in large part to FTX's ven-ture investment in Anthropic, the high-flying AI company).

Just weeks before the fraud was uncovered, Sequoia published a gushing 13,000-word essay extolling Bankman-Fried's genius and describing the research Sequoia's partners did to understand the business. "The exchange that [Bankman-Fried] had started to build, FTX, was Goldilocks-perfect," the essay states. "There was no concerted effort to skirt the law, no Zuckerbergian diktat demanding that things be broken." It then went on to quote the Sequoia associate responsible for the FTX investment: "[Bankman-Fried] is committed to making the right chess moves for FTX to eventually be able to legally do everything they want to do in the US . . . not by asking forgiveness, but by asking permission." When FTX collapsed, Sequoia showed very little remorse, telling its limited partners casually, "Some investments will surprise to the upside and some will surprise to the downside."

The SEC wasn't so sanguine. The agency, under Gary Gensler's leadership, has been exploring ways to hold venture capitalists to account, and the FTX saga added fuel to his efforts. The SEC had been working on rules that would require venture firms and other private market investors to be more transparent with their LPs, allowing LPs to more easily sue venture capitalists for malfeasance. In summer 2023 those rules were finalized. Though the final version stripped out the provisions that would have made it easier for LPs to sue fund managers, the transparency requirements remained. But even those were a bridge too far for investors. The new rules would have required fund managers to report on their activities, particularly the fees they are charging, on a quarterly basis. They also would have been required to perform annual audits on all the funds they manage and make those audit reports available to LPs.

Private fund managers were apoplectic about the rule changes. Bobby Franklin, the CEO of the NVCA at the time, told the *Financial Times* that the rules had the "potential to stifle innovation and harm the economic environment for venture and start-ups." Shortly after the rules were finalized, the NVCA joined other industry groups to sue the SEC to have the rules invalidated. In June 2024, a federal appeals court in Louisiana sided with the investors, striking down the SEC's newly asserted authority.

Though it may be the most brazen, the action perpetrated by Bankman-Fried and FTX is far from the only fraud committed by startup founders. This fraudulent activity is only possible because of the investors who not only give the fraudsters the fuel to build their schemes but also employ the VC hype machines to ensure the founders have the most credibility and the furthest reach possible. Allowing them to claim ignorance is an unacceptable injustice. Even if, as investors often argue when frauds like this are uncovered, the investors were kept in the dark about illegal activity, they are still in a position to conduct rigorous diligence that would identify fraudulent activity before investments are made. Instead of looking the other way, they should be forced to require that companies provide more insight into their finances and activities even *after* investments are made.

Investors will argue that holding them accountable for the bad behavior of the founders they invest in will have a chilling effect on innovation. If they have to play it too safe, the economy will miss out on the big winners that venture capital creates. That argument points to exactly why the venture capital model is harmful. If an innovation ecosystem hinges on its ability to let a few frauds slip through the cracks, it's probably a sign that there is something rotten at the core.

Historically, the venture capital industry has been able to avoid coming under too much scrutiny from the SEC and other regulators because, its proponents have argued, it isn't "systemically important" enough to the overall economy to warrant robust oversight.

Unlike banking, the argument goes, venture capital doesn't control enough economic activity to take anything else down with it if it gets too far off the rails. But while it might not be as systemically important as investment banking, it certainly holds outsized sway in the economy overall. Almost half of all the companies that went public between 1995 and 2019 were venture backed, even though less than 0.5 percent of companies receive venture capital. The venture-backed companies in that cohort accounted for more than 75 percent of the total market cap by 2019.

We've also seen high-profile examples of the ability venture capitalists have to influence broader markets when they drove the collapse of Silicon Valley Bank in 2022, almost causing another banking crisis as a result. Beyond the financial system, venture capital's influence on all aspects of our economy—including ones I've laid out in this book—demonstrates its systemic importance to the economic health of the country. Given how central venture capital is to the country's future economic growth, it may be time to consider whether there are more stringent and/or public reporting requirements that make sense. These might apply only to larger funds, much as Dodd-Frank's reporting requirements apply only to banks of a certain size. This would also create an additional incentive to keep funds small.

It isn't just the funds that should be subject to increased reporting. Privately held companies are not required to abide by the same accounting rules that apply to public companies, for the same reasoning that applies to private funds. But that thinking no longer applies in an era when private companies can be just as large, or even larger, than publicly traded companies. There has been a proliferation of unicorns over the last decade as more venture-backed startups decided to stay private for longer periods of time, sometimes exactly so they can avoid the scrutiny that public companies endure. There is a growing consensus that these companies should have to be more transparent given their size and influence in the economy, and the SEC is said to be considering new rules that would require private companies valued at more than $1 billion to submit to reporting requirements similar to those of public companies.

While all these solutions are simple, none of them are easy. The stranglehold that the power law approach to venture investing has on our financial system is deeply embedded, not just in practice but in the identities of the various stakeholders who participate in the system. Those who are winning in the current system—the ecosystem of fund managers, lawyers, and consultants who fight to maintain the status quo—are enormously powerful. But the people who stand to gain the most—the entrepreneurs, pension holders, those who believe in the nonprofit mission of cultural, academic, and philanthropic institutions—also have collective power. If we are to advance a more positive vision for America's innovation ecosystem, all of us will need to step off the sidelines.

CONCLUSION

When Georges Doriot started American Research and Development in the 1940s, he had no way of knowing what would grow from the seed he was planting. His goal, along with the other businessmen of his time who were worried about where economy-spurring innovation would come from, was to create a space where risk-taking could be rewarded and where the intrepid spirit of American entrepreneurship could flourish. It didn't come easy, and he had his fair share of doubters, but his early investments, and the pioneering investment firms he inspired, enabled many of the technological breakthroughs we all take for granted today.

Corrupted by the sheer amount of money flowing to the industry and a raft of speculators who see tech as a get-rich-quick opportunity, the methodology that worked for Doriot has moved well beyond its most useful purpose. Far from being the home to experimental, countercultural approaches to fostering innovation that it was in the 1960s and 1970s, Silicon Valley has become a corporatist monoculture that enforces conformity in order to reap profits.

That forced conformity is hurting the rest of us in two ways. *First*, the blind desperation with which venture capitalists pursue power law returns elevates and enables fraudsters, liars, and out-right criminals. Because so much money is chasing so few real op-portunities that have the potential to deliver venture-scale returns, venture capitalists adopt a level of credulity that shameless pitch people like Adam Neumann and Sam Bankman-Fried are more than happy to exploit. The better they are at playing along, the more money flows their way. And entrepreneurs who operate with more scruples get the message that in order to play the game, one must push the envelope. The rest of us often end up paying the price.

Second and, I think, more important, the venture capital mono-culture prevents good businesses—many of which could deliver break-throughs on tough problems like the housing and climate crises—from thriving. The tragic paradox of mainstream venture capital, I have come to believe, is that as its practitioners become more des-perate to achieve power law returns, they are killing more innova-tion than they create. VCs are dumping billions into companies that excel at marketing and can tell a good story about growth but aren't building anything particularly innovative. (Has anybody asked why mattress companies or fast casual restaurants need high-risk ven-ture capital investments?) This is making the economy less vibrant. As Rana Foroohar points out in her 2016 book, *Makers and Takers: How Wall Street Destroyed Main Street*, the number of small businesses created since the 1980s has fallen every year, despite the invention of so many new technologies. That data includes companies of all kinds, but it is striking that the reduction in new business activity persists, even despite the advent of the internet and the rise of digital technology.

In light of this, why are so many smart capitalists leaving so much money on the table? The best answer I can come up with is that capitalism isn't nearly as rational as some like to think. The deep entrenchment of the power law approach reflects a herd mentality that has been hardened over decades into an orthodoxy that few feel is in their interest to challenge. Rather than take a methodical approach to building the innovation economy, fortune hunters have started trying to replicate only the return profiles of the early, most successful pioneering funds, and not their approach to identifying breakthrough technological innovations. Meanwhile, limited partners have been all too eager to support those efforts.

At the same time, the people who are pursuing a different approach aren't getting the attention and support they would need to prove a different path is possible. Over time, the care and feeding of one model at the expense of any other solidified into the ideological commitment to the power law approach we see in venture capital today. The sometimes toxic male-dominant culture of Silicon Valley, where bigger is always better, also plays a role. As one startup founder put it to me, the strategy of seeking returns by hitting doubles rather than grand slams is seen as too "beta" by the alpha male ecosystem of venture capital.

Many people have asked me over the course of my time writing this book, "Aren't you just describing capitalism?" Maybe they're right and venture capital is just the most egregious manifestation of the casino capitalism that seems to be all around us. Late-stage American capitalism's id. But I don't think that explanation is comprehensive enough. We hold the tech industry to a different standard than, say, finance or the fossil fuel industry, because tech represents a forward-facing vision, allowing us to imagine a future that is better than the past. That future looks pretty bleak at the moment; a venture capital system reoriented around fueling breakthrough technologies that advance human flourishing will help us reclaim our optimism. At its best, venture capital is a key that unlocks massive net benefit for humankind, and there are few times in history that have called for it more than now.

A s I'm writing this book, the venture industry is in a lull. The era of easy money, when VCs were happy to subsidize money-losing businesses indefinitely, seems to be over for the time being. As interest rates have risen, limited partners are driving a harder bargain, asking more from VCs and the companies they invest in. While founders have been used to raising venture capital by telling stories of scale, now they are being forced to figure out how they will turn a profit.

These dynamics, where the balance of power has swung to investors and when companies that demonstrate solid business fundamentals are suddenly great investment opportunities, won't last forever. If there is one thing you can be confident about in the world of venture capital, it's the cyclical nature of the ecosystem. The breathtaking speed with which generative artificial intelligence has come on the scene, and the clamor it has created among investors to get a foothold in the best deals, indicates the next boom cycle is already underway. This short moment of relative austerity offers an opportunity for venture capital to reconstitute itself to its appropriate role in the economy: finding and funding the very small number of businesses that will deliver the breakthrough technologies that can help us tackle the biggest challenges humanity faces—and yes, also deliver the huge financial returns that come with those kinds of breakthroughs. It's also an opportunity for all of us to design different options that can support the high-growth businesses that can change our lives for the better but don't need to become hyperscale in order to do so. We need a more diverse ecosystem of risk capital and investors who are willing to support the pursuit of different return generation models. We need to foster the emergence of a new asset class: high-growth startups that play a critical role in growing the economy but that don't require hyper maximalist returns in order to get there.

To catalyze this change, some of you reading this have some reflecting to do on the role you play in the current system. All of us have some culpability; some of us have more power to effect change than others.

Limited partners, especially those in private philanthropy who are particularly interested in the future the tech industry is creating, must act now. They must use their resources to seed a more robust ecosystem of Indie-style investors, and their platforms and positionality to change the hearts and minds of other limited partners. There are few more effective ways to deploy resources—and a relatively small amount at that—to fundamentally institute change that advances their missions.

For policymakers, and those of us who advocate for public policy: as we consider how to address the harms we attribute to the tech industry, we must widen our aperture to account for the role of venture capital. Too many of us have ignored the economic dynamics that shape the way technology is designed and distributed. Our narrow vision doesn't just make it harder for us to rein in Big Tech, it risks stifling the innovation that may ultimately solve some of the biggest challenges facing our planet.

And finally, I speak directly to the entrepreneurs, especially the ones I talked to while researching this book: Don't give up. Don't listen to investors who tell you your ideas aren't big enough or who try to shake your conviction in the solutions you are building. Don't be afraid to forge a different path, taking heart from the experiences of others, like Thompson Aderinkomi and Mike Salguero, that the rewards can be just as great—if not greater—by taking the road less traveled. We need your ingenuity and bravery now more than ever.

ACKNOWLEDGMENTS

Writing a book truly has been the most challenging experience of my life, one that I can't imagine completing without the kindness, grace, and encouragement of an amazing set of people. Cassidy and Leila picked me up off the mat at my lowest point, when I had convinced myself I couldn't do this. Their steady hands and constant enthusiasm were the difference between the book you're holding and a bunch of abandoned thoughts scattered across a set of Google Docs. The staff and board at TechEquity allowed me to take the time I needed away from running the organization at a critical juncture in its development, picking up the slack in my absence and making it better off than I left it. Mentors like John Palfrey and Colin Maclay exuded steadfast confidence on my behalf, even (especially) when I didn't have much myself. My fellow book warriors Jen Pahlka and Natalie Foster taught me so much by example as they produced magnificent and inspiring books themselves. I'm not sure what I would have done without them showing me there was a light at the end of the tunnel. The Bougie crew, the best village a gal could ask for, always lent sympathetic ears to my constant complaining. If it annoyed them, they never showed it!

The process of researching and writing this book was a learning journey, and I'm grateful to Roy Bahat and Elana Berkowitz for being patient teachers about the practice of VC. I'm of course also massively grateful to the dozens of entrepreneurs and investors—some whose voices appear in this book and many more whose do not—who shared their experiences with me. They were often painful recountings, and I'm grateful they trusted me with them. I hope I did you justice. I'm also grateful to Micah Sifry who gave me wonderfully constructive feedback on early drafts and, through his own clear and morally resonant writing on tech and democracy, provided valuable inspiration.

Speaking of learning, I was always a curious kid, but I gained an understanding that the pursuit of my curiosities was divine from the Jesuits at Boston College (go Eagles!) who opened my eyes to the wonder and beauty inherent in knowledge.

My greatest teachers have always been my family: my parents, who taught me kindness and shaped my sense of justice; my siblings, who always kept me humble; and my nieces and nephew, who taught me how selfless love could be.

Just after I completed this book I lost my constant companion, my dog, Gordon. He gifted me the unconditional love I needed to get through this process, in the way only a dog can. He was by my side through some of the toughest moments of my life and, I like to think, left me just when he knew I'd be okay on my own. Thank you, Gordy, my angel. See you on the other side.

NOTES

INTRODUCTION

2 **inequality hit unprecedented levels:** Naomi Zewde and Stephen Crystal, "Impact of the 2008 Recession on Wealth-Adjusted Income and Inequality for U.S. Cohorts," *Journals of Gerontology B, Psychological Sciences and Social Sciences* 77, no. 4 (2022): 780–89, https://www.ncbi.nlm.nih.gov/pmc/articles/PMC9122750.

2 **Rents in the Bay Area grew:** Julie Littman, "Rents in Bay Area Major Metros Increased by about 50% since 2010," *Bisnow*, April 3, 2018, https://www.bisnow.com/san-francisco/news/multifamily/rents-in-bay-area-major-metros-increased-by-about-50-since-2010-86847.

2 **boneheaded, insensitive things:** Brock Keeling, "Startup CEO Trashes SF's Homeless 'Degenerates,'" Sfist.com, December 11, 2013, https://sfist.com/2013/12/11/angelhack_startup_ceo_trashes_sfs_d.

8 **"enshrin[e] strong antitrust principles":** Elizabeth Warren, "Warren Delivers Remarks at Freedom from Facebook and Google: Break Up Big Tech," press release, May 27, 2021, https://www.warren.senate.gov/newsroom/press-releases/warren-delivers-remarks-at-freedom-from-facebook-and-google-break-up-big-tech.

8 **"competition is for losers":** Peter Thiel, "Competition Is for Losers," *Wall Street Journal*, September 12, 2014, https://www.wsj.com/articles/peter-thiel-competition-is-for-losers-1410535536.

11 **With an estimated 40 percent:** J. C. Reindl, "Why Detroit's Lights Went Out," *Detroit Free Press*, November 17, 2013, https://www.usatoday.com/story/news/nation/2013/11/17/detroit-finances-dark-streetlights/3622205.

18 **"More and more major businesses"**: Marc Andreessen, "Why Software Is Eating the World," a16z.com, August 20, 2011, https://a16z.com/2011/08/20/why-software-is-eating-the-world.

20 **"The expectation for what a good company is"**: Berber Jin and Nate Rattner, "High Interest Rates Crushed Startup Investment. Here Is What Could Revive It," *Wall Street Journal*, January 28, 2024, https://www.wsj.com/finance/investing/high-interest-rates-crushing-startup-investments-cc76f82f.

20 **A survey by the Kauffman Fellows**: "Kauffman Fellows Sentiment Survey 2024," https://www.kauffmanfellows.org/resources/sentiment-survey.

1

THE METHODOLOGY

21 **"Venture capital is not even"**: Bill Gurley in Tren Griffin, *A Dozen Lessons for Entrepreneurs* (New York: Columbia Business School Publishing, 2017).

24 **Breakthrough entrepreneurs in the early part of the twentieth century**: Martin Kenney, "How Venture Capital Became a Component of the U.S. National System of Innovation," *Industrial and Corporate Change* 20, no. 6 (2011): 1682, https://kenney.faculty.ucdavis.edu/wp-content/uploads/sites/332/2018/03/how-venture-capital-became-a-component-of.pdf.

24 **Thomas Edison and Henry Ford**: Jason Crawford, "How Early American Inventors Funded Their Ventures," RootsofProgress.org, March 8, 2020; "The Life of Henry Ford," *American Experience*, PBS, https://www.pbs.org/wgbh/americanexperience/features/henryford/#:~:text=Henry%20Ford%20and%20his%20partner,or%20business%20contacts%20of%20Malcomson.

25 **"It is very risky to put money"**: Merrill Griswold quoted in Tom Nicholas, *VC: An American History* (Cambridge, MA: Harvard University Press, 2019), 115.

25 **Without DEC, ARD wouldn't have outperformed**: Nicholas, *VC: An American History*, 131.

27 **Perhaps the biggest advantage**: "Private Fund," U.S. Securities and Exchange Commission, modified April 25, 2024, https://www.sec.gov/education/capitalraising/building-blocks/private-fund.

29 **their rise in popularity in the mid-twentieth century**: Nicholas, *VC: An American History*, 151–55.

31 **not enough investors were willing:** Molly Sauter, "A Businessman's Risk: The Construction of Venture Capital at the Center of U.S. High Technology" (PhD Thesis, McGill University, 2021), https://escholarship.mcgill.ca /concern/theses/j6731853z.

32 **In 1978, venture capital funds were managing:** Paul A. Gompers, "The Rise and Fall of Venture Capital," *Business and Economic History* 23, no. 2 (1994), https://thebhc.org/sites/default/files/beh/BEHprint/v023n2/p0001 -p0026.pdf.

33 **lower the top marginal capital gains tax rate:** The capital gains tax rate has since been lowered even further; the top marginal rate stands at 20 percent as of this writing.

33 **Inequality has skyrocketed and wages:** Juliana Menasce Horowitz, Ruth Igielnik, and Rakesh Kochhar, "Trends in Income and Wealth Inequality," Pew Research Center, 2020, https://www.pewresearch.org/social-trends /2020/01/09/trends-in-income-and-wealth-inequality.

33 **NVCA continues to fight hard to maintain low tax rates for investment income:** Alan Rappeport, Emily Flitter, and Kate Kelly, "The Carried Interest Loophole Survives Another Political Battle," *New York Times*, August 5, 2020, https://www.nytimes.com/2022/08/05/business/carried-interest -senate-bill.html; Jeff Farrah, "Venture Industry's 2022 Policy Priorities," NVCA, https://nvca.org/venture-industrys-2022-policy-priorities.

34 **more than $128 billion was raised:** Cameron Stanfill, Kyle Stanford, and Joshua Chao, "Q4 2021 PitchBook-NVCA Venture Monitor First Look," PitchBook, January 5, 2022, https://files.pitchbook.com/website/files/pdf /Q4_2021_PitchBook_NVCA_Venture_Monitor_First_Look.pdf.

35 **"Great venture capitalists and venture capital funds":** Chip Hazard, "Venture Investor's Playbook—Part 1: 4 Success Factors for Early-Stage Venture Investors," LinkedIn, September 17, 2019, https://www.linkedin.com/pulse /venture-investors-playbook-part-1-chip-hazard/.

35 **Mallaby describes the returns of Horsley Bridge:** Sebastian Mallaby, *The Power Law* (New York: Penguin Press, 2022), 8.

35 **just 6 percent of them accounted for 60 percent:** Benedict Evans, "In Praise of Failure," Ben-Evans.com, August 10, 2016, https://www.ben-evans .com/benedictevans/2016/4/28/winning-and-losing.

36 **"invest in a particular type of startup":** Evans, "In Praise of Failure."

37 **"it's just not possible to make the single and double concept work":** Tracy Alloway and Joe Weisenthal, "Transcript: Jason Calacanis on the Expensive Lesson Coming to Silicon Valley," Bloomberg.com, July 21, 2022, https://

www.bloomberg.com/news/articles/2022-07-21/transcript-jason-calacanis -on-the-expensive-lesson-coming-to-silicon-valley.

38 **"I sell jet fuel":** Erin Griffith, "More Start-Ups Have an Unfamiliar Message for Venture Capitalists: Get Lost," *New York Times*, January 11, 2019, https:// www.nytimes.com/2019/01/11/technology/start-ups-rejecting-venture -capital.html.

39 **no better articulation:** Reid Hoffman and Chris Yeh, *Blitzscaling* (New York: Currency, 2018).

39 **Everything you need to know about** *Blitzscaling*: Tim O'Reilly, "The Fundamental Problem with Silicon Valley's Favorite Growth Strategy," *Quartz*, February 5, 2019, https://qz.com/1540608/the-problem-with-silicon-valleys -obsession-with-blitzscaling-growth.

40 **"an accelerant that allows your company to grow":** Hoffman and Yeh, *Blitzscaling*, 12.

40 **"anyone who wants to understand":** Hoffman and Yeh, *Blitzscaling*, 19.

40 **"embrace counterintuitive rules":** Hoffman and Yeh, *Blitzscaling*, 48.

41 **"Uber often uses heavy subsidies":** Hoffman and Yeh, *Blitzscaling*, 47.

41 **Its Blitzscaling approach drove it to steamroll:** Ben Butler, "The Uber Files: Firm Knew It Launched Illegally in Australia, Then Leaned on Governments to Change the Law," *The Guardian*, July 14, 2022, https://www.theguardian .com/news/2022/jul/15/the-uber-files-australia-launched-operated-illegally -document-leak.

41 **In** *Super Pumped*: Mike Isaac, *Super Pumped: The Battle for Uber* (New York: W. W. Norton, 2020).

41 **a program at Uber called Greyball:** Mike Isaac, "How Uber Deceives the Authorities Worldwide," *New York Times*, March 3, 2017, https://www .nytimes.com/2017/03/03/technology/uber-greyball-program-evade -authorities.html.

42 **enlisted employees to book, and then promptly cancel:** Cassandra Khaw, "Uber Accused of Booking 5,560 Fake Lyft Rides," *The Verge*, August 12, 2014, https://www.theverge.com/2014/8/12/5994077/uber-cancellation -accusations.

42 **"We're just fucking illegal":** "The Uber Files," International Consortium of Investigative Journalists, https://www.icij.org/investigations/uber -files/.

42 **a former Uber software engineer:** Susan Fowler, "Reflecting on One Very, Very Strange Year at Uber," susanjfowler.com, February 19, 2017, https:// www.susanjfowler.com/blog/2017/2/19/reflecting-on-one-very-strange -year-at-uber.

43 **rise of reports of sexual assaults:** Sam Levin, "Uber Accused of Silencing Women Who Claim Sexual Assault by Drivers," *The Guardian*, March 15, 2018, https://www.theguardian.com/technology/2018/mar/15/uber-class -action-lawsuit-sexual-assault-rape-arbitration.

45 **amount of money invested by LPs:** Pitchbook, NVCA, https://nvca.org/wp -content/uploads/2024/04/2024-YB-data-PDF.pdf.

45 **In 2018, megafunds raised 44 percent:** "Global Private Market Fundraising Report," PitchBook, updated October 5, 2023, https://pitchbook.com/news /reports/q2-2023-global-private-market-fundraising-report.

45 **Armstrong did the math on one megafund:** Evan Armstrong, "Venture Capital Is Ripe for Disruption," Every.to, October 27, 2022, https://every.to /napkin-math/venture-capital-is-ripe-for-disruption.

46 **made somewhere between a 30x and 90x return:** Kate Clark, "The Biggest VC Winners from Figma's $20 Billion Sale," *The Information*, September 15, 2022, https://www.theinformation.com/articles/the-biggest-vc-winners -from-figmas-20-billion-sale?rc=1laxcc.

47 **"Groups that do not increase their size":** Victoria Ivashina and Josh Lerner, *Patient Capital: The Challenges and Promises of Long-Term Investing* (Princeton, NJ: Princeton University Press, 2019), https://press.princeton.edu/books /hardcover/9780691186733/patient-capital.

48 **firms like Insight Partners:** Tabby Kinder and George Hammond, "Insight Partners Cuts Size of $20bn Fund Amidst 'Great Reset in Tech,'" *Financial Times*, June 12, 2023, https://www.ft.com/content/cccc80cc-efe0-41c2 -bf27-54c69bac28d4.

48 **and Tiger Global:** Hema Parmer, "Tiger Global VC Fund Closes 63% Below Target with $2.2 Billion," Bloomberg.com, April 1, 2024, https://www .bloomberg.com/news/articles/2024-04-01/tiger-global-vc-fund-closes -63-below-target-with-2-2-billion.

51 **Less than 2 percent of clean energy companies:** Juliet Eilperin, "Why the Clean Tech Boom Went Bust," *Wired*, January 20, 2012, https://www.wired .com/2012/01/ff_solyndra.

51 **longer time frames for venture funds might be necessary:** DSN, "Weathering the Storm: Kleiner Perkins and the Tragedy of Clean-Tech Venture Capital," Harvard Business School, HBS Digital Initiative, MBA Student Perspectives, November 4, 2016, https://d3.harvard.edu/platform-rctom /submission/weathering-the-storm-kleiner-perkins-and-the-tragedy-of -clean-tech-venture-capital/#_ftn6.

2
THE FOUNDERS

57 **A key part of YC's gene-splicing curriculum:** Steven Levy, "How Y Combinator Changed the World," *Wired*, December 21, 2021, https://www.wired.com/story/how-y-combinator-changed-the-world.

69 **Good Eggs warehouse and delivery workforce:** "We believe Good Food is the most powerful force for change," GoodEggs.com, https://careers.goodeggs.com.

76 **"Shefs are aunties and abuelas":** "The Power of a Homemade Meal," Shef blog, https://blog.shef.com/our-story.

77 **celebrities like Katy Perry and Odell Beckham Jr.:** Amy Feldman, ed., "The Next Billion-Dollar Startups 2021," *Forbes*, October 12, 2021, https://www.forbes.com/sites/amyfeldman/2021/10/12/next-billion-dollar-startups-2021/?sh=65c00f41f1cd.

78 **"varies in each market":** "Make Money Doing What You Love," Shef.com, https://shef.com/become-a-shef.

80 **"Each Seller is solely responsible":** Shef Terms of Service, Shef.com, revised December 16, 2022, https://shef.com/terms-of-service.

80 **Shef's terms of service make it clear:** "The Health + Safety of Our Community Will Always Come First," Shef.com, https://shef.com/homemade-food/food-safety.

81 **"I am fairly new to the venture":** "Can Someone Share Their Experience with Cooking for Shef," Reddit, 2022, https://www.reddit.com/r/Shef/comments/vclktm/can_someone_share_their_experience_with_cooking.

82 **"When any other person had an idea":** Ron Lieber, "How Charlie Javice Got JPMorgan to Pay $175 Million for . . . What Exactly?" *New York Times*, January 21, 2023, https://www.nytimes.com/2023/01/21/business/jpmorgan-chase-charlie-javice-fraud.html.

3
VENTURE CAPITAL AND LABOR

87 **up to 30 percent of US employers have misclassified employees:** Lynn Rhinehart, Celine McNicholas, Margaret Poydock, and Ihna Mangundayao, "Misclassification, the ABC Test, and Employee Status," Economic Policy

Institute, June 16, 2021, https://www.epi.org/publication/misclassification
-the-abc-test-and-employee-status-the-california-experience-and-its
-relevance-to-current-policy-debates.

92 **"online marketplaces avoid many"**: Reid Hoffman and Chris Yeh, *Blitzscaling*
(New York: Currency, 2018), 87.

93 **statewide ballot initiative**: Ken Jacobs and Michael Reich, "The Effects
of Proposition 22 on Driver Earnings: Response to a Lyft-Funded Report by
Dr. Christopher Thornberg," UC Berkeley Labor Center, August 26, 2020,
https://laborcenter.berkeley.edu/the-effects-of-proposition-22-on-driver
-earnings-response-to-a-lyft-funded-report-by-dr-christopher-thornberg.

96 **"a way forward for redefining the relationship"**: Shawn Carolan, "What
Proposition 22 Now Makes Possible," *The Information*, November 10, 2020,
https://www.theinformation.com/articles/what-proposition-22-now-makes
-possible.

99 **data collected by the Alphabet Workers Union**: "Every Google Worker,"
Alphabet Workers Union, undated, https://everygoogleworker.alphabet
workersunion.org.

99 **"the Microsoft case provides guidance"**: Erin Hatton, *The Temp Economy:
From Kelly Girls to Permatemps in Postwar America* (Philadelphia: Temple University Press, 2011), 123.

100 **Google had 121,000 temp workers**: Daisuke Wakabayashi, "Google's Shadow
Workforce: Temps Who Outnumber Full-Time Employees," *New York Times*,
May 28, 2019, https://www.nytimes.com/2019/05/28/technology/google
-temp-workers.html.

104 **they create firewalls for temps and contractors**: These firewalls are often much more stringent than the law requires, but tech companies take
the most conservative interpretation in order to avoid any appearance of
co-employment.

106 **she found herself unable to access her email**: Carolyn Said, "Proposed
California Bill Would Expand Protection of Laid-Off Workers," *San Francisco Chronicle*, March 7, 2023, https://www.sfchronicle.com/tech/article
/california-bill-layoff-workers-17819130.php.

106 **contractor instead of a full-time employee**: Lauren Rosenblatt, "Laid Off
by Big Tech, Then Recruited for Contract Work—at the Same Place," *The
Seattle Times*, April 16, 2023, https://www.seattletimes.com/business/laid
-off-by-big-tech-then-recruited-for-contract-work-at-the-same-place.

106 **gender, racial, and ethnic minority groups are overrepresented**: *Shining a
Light on Tech's Shadow Workforce*, Contract Worker Disparity Project—2022

Report, TechEquity Collaborative, https://contractwork.techequitycollabo rative.org.

110 **Deloitte found zero Black or Latino investment partners:** *VC Human Capital Survey*, 4th ed., April 2023, Deloitte, https://www2.deloitte.com/us/en /pages/audit/articles/diversity-venture-capital-human-capital-survey.html.

110 **the social connections:** Tom Nicholas, *VC: An American History* (Cambridge, MA: Harvard University Press, 2019), 161–71.

112 **As venture capital investment constricted:** Courtney Connley, "Black and Latinx Founders Have Received Just 2.6% of VC Funding So Far in 2020, according to New Report," CNBC, October 8, 2020, https://www.cnbc.com /2020/10/07/black-and-latinx-founders-have-received-just-2point6percent -of-vc-funding-in-2020-so-far.html.

112 **funding to startups led by Black founders:** Gabrielle Fonrouge, "Venture Capital for Black Entrepreneurs Plummeted 45% in 2022, Data Shows," February 2, 2023, CNBC, https://www.cnbc.com/2023/02/02/venture -capital-black-founders-plummeted.html.

112 **"the right kind of nerdy":** NPR staff, "Failure: The F-word Silicon Valley Loves and Hates," *All Things Considered*, NPR, June 19, 2012, https://www .npr.org/2012/06/19/155005546/failure-the-f-word-silicon-valley-loves -and-hates.

113 **"Black entrepreneurs don't need":** Nico Grant, "Black Venture Capitalists Confront Silicon Valley's Quiet Racism," Bloomberg.com, August 24, 2020, https://www.bloomberg.com/news/features/2020-08-24/black-venture -capitalists-confront-silicon-valley-s-quiet-racism.

114 **When companies are growing fast:** "Diversity in High Tech," U.S. Equal Employment Opportunity Commission, undated, https://www.eeoc.gov /special-report/diversity-high-tech.

115 **Google's Black and Latino workforce:** Rupert Neate, "Facebook Only Hired Seven Black People in Latest Diversity Count," *The Guardian*, June 25, 2015, https://www.theguardian.com/technology/2015/jun/25/facebook-diversity -report-black-white-women-employees.

116 **25 percent of underrepresented people of color:** Allison Scott, Freada Kapor Klein, and Uriridiakoghene Onovakpuri, *Tech Leavers Study*, Kapor Center, April 27, 2017, https://www.kaporcenter.org/the-2017-tech-leavers -study/.

116 **Diversity and inclusion roles were among the most affected:** Kelsey Butler, "Big Tech Layoffs Are Hitting Diversity and Inclusion Jobs Hard," Bloomberg.com, January 24, 2023, https://www.bloomberg.com/news

/articles/2023-01-24/tech-layoffs-are-hitting-diversity-and-inclusion-jobs
-hard.

116 **lack of diversity in the tech industry:** Kate Wittels, Shuprotim Bhaumik,
Bret Nolan Collazzi, and Ashley So, "Tech's Racial Equity Gap Costs Black
& Latinx Workers $50B a Year. Our Collaborators Prescribe Solutions,"
HR&A Advisors, April 23, 2021, https://www.hraadvisors.com/techs-racial
-equity-gap-costs-black-latinx-workers-50b-a-year-our-collaborators
-prescribe-solutions.

4

VENTURE CAPITAL AND HOUSING

119 **Scholars like Richard Rothstein and Keeanga-Yamahtta Taylor:** Richard
Rothstein, Distinguished Fellow, Economic Policy Institute, https://www.epi
.org/people/richard-rothstein/; Keeanga-Yamahtta Taylor, *Race for Profit*
(Chapel Hill: University of North Carolina Press, 2020).

120 **eight out of ten American metropolitan areas:** Stephen Menendian, Samir
Gambhir, and Arthur Gailes, *The Roots of Structural Racism Report*, Other-
ing and Belonging Institute, June 21, 2021, updated June 30, 2021, https://
belonging.berkeley.edu/roots-structural-racism.

120 **gap between white homeownership and Black homeownership:** Courtney
Connley, "Why the Homeownership Gap between White and Black Americans Is Larger Today Than It Was over 50 Years Ago," CNBC, August 21,
2020, https://www.cnbc.com/2020/08/21/why-the-homeownership-gap
-between-white-and-black-americans-is-larger-today-than-it-was-over
-50-years-ago.html.

120 **more than half of all foreclosed properties:** Jacob W. Faber, "Racial Dynamics of Subprime Mortgage Lending at the Peak," NYU Scholars, April 2013,
https://nyuscholars.nyu.edu/en/publications/racial-dynamics-of-subprime
-mortgage-lending-at-the-peak; "Black and Hispanic Communities Are Still
Reeling from the Foreclosure Crisis," Zillow, April 25, 2019, https://zillow
.mediaroom.com/2019-04-25-Black-and-Hispanic-Communities-are-Still
-Reeling-from-the-Foreclosure-Crisis.

121 **Six million American families lost their home:** "Joint Statement by Inde-
pendent United Nations Human Rights Experts Warning of the Threat That
Financial Speculation Poses to the Enjoyment of a Range of Human Rights,"
United Nations, Office of the High Commissioner, Human Rights, October

19, 2021, https://www.ohchr.org/en/statements/2021/10/joint-statement
-independent-united-nations-human-rights-experts-warning-threat.

122 **share of households headed by renters:** Anthony Cilluffo, A. W. Geiger, and Richard Fry, "More U.S. Households Are Renting Than at Any Point in 50 Years," Pew Research Center, July 19, 2017, https://www.pewresearch .org/fact-tank/2017/07/19/more-u-s-households-are-renting-than-at-any -point-in-50-years.

122 **Wall Street's interest in real estate:** Alana Semuels, "When Wall Street Is Your Landlord," *The Atlantic*, February 13, 2019, https://www.theatlantic .com/technology/archive/2019/02/single-family-landlords-wall-street /582394.

123 **"large-scale SFR ownership and management":** Doug Brien and Colin Wiel, *The Big Long* (Herndon, VA: Amplify Publishing, 2022), 13.

124 **buy foreclosed properties in bulk:** Brien and Wiel, *The Big Long*, 49.

124 **"Technology has disrupted":** Semuels, "When Wall Street Is Your Land-lord."

124 **the Kafkaesque horror:** Francesca Wari, "A $60 Billion Housing Grab by Wall Street," *New York Times*, March 4, 2020, updated October 22, 2021, https://www.nytimes.com/2020/03/04/magazine/wall-street-landlords .html.

125 **increased its fee income by 114 percent:** Semuels, "When Wall Street Is Your Landlord."

125 **pattern of blatant disregard:** "In the Wake of the Great Recession, Investors Have Scooped Up Thousands of Single-Family Homes across Metro At-lanta," *Atlanta Journal-Constitution*, February 16, 2023, https://www.ajc .com/american-dream/renter-evictions/.

125 **young children exposed to mold:** Wari, "A $60 Billion Housing Grab by Wall Street."

125 **sewage overflowing into homes:** Semuels, "When Wall Street Is Your Landlord."

125 **much more likely to initiate eviction proceedings:** Elora Raymond, Rich-ard Duckworth, Ben Miller, Michael Lucas, and Shiraj Pokharel, "Corpo-rate Landlords, Institutional Investors, and Displacement: Eviction Rates in Single-Family Rentals," Federal Reserve Bank of Atlanta, Community & Economic Development Discussion Paper, no. 04-16, December 2016, https:// www.atlantafed.org/-/media/documents/community-development/publica tions/discussion-papers/2016/04-corporate-landlords-institutional-investors -and-displacement-2016-12-21.pdf.

126 **worldwide "property tech" market, or Proptech:** "Value of venture capital

investments in proptech companies worldwide from 2008 to 1st half 2022," Statista Research Department, February 20, 2023, https://www.statista .com/statistics/1060470/global-proptech-venture-capital-investment-value.

127 **largest residential institutional owners:** "We Are Backed by Technology-Forward Corporations from Every Corner of the Built World," Fifth Wall, undated, retrieved February 23, 2024, https://fifthwall.com/partners.

127 **"dramatically accelerate the growth":** "Introducing Fifth Wall Ventures," Medium.com, May 1, 2017, https://medium.com/fifth-wall-insights/intro ducing-fifth-wall-ventures-c806bfa33082.

127 **"end-to-end real estate platform":** "Your One Stop Shop to . . . ," MYND, undated, retrieved August 4, 2023, https://www.mynd.co.

128 **"extract optimal value":** Brien and Wiel, *The Big Long*, 155.

128 **"impressed by Mynd's early progress":** "Mynd Property Management Raises $5.1 Million in Series A1 Financing," PR Newswire, March 21, 2017, https://www.prnewswire.com/news-releases/mynd-property-management -raises-51-million-in-series-a1-financing-300426746.html.

128 **Mynd has sinced raised:** Mary Azevedo, "Mynd Raises $57.3M at an $807M Valuation to Give People a Way to Invest in Rental Properties Remotely," *TechCrunch*, September 6, 2021, https://techcrunch.com/2021/09/16/mynd -raises-57-3m-from-qed-investors-at-a-870m-valuation-to-give-people -a-way-to-invest-in-rental-properties-remotely.

129 **Invesco, the behemoth investment firm:** Greg Isaacson, "Invesco Forms $5B SFR Partnership with Mynd," *Multi-Housing News*, June 2, 2021, https:// www.multihousingnews.com/invesco-forms-5b-sfr-partnership-with -mynd.

129 **half of the homes sold in 2021 go to private equity firms:** *DeKalb County Housing Affordability Study*, March 2022, https://www.scribd.com/document /619370298/DeKalb-County-Housing-Affordability-Study.

130 **drop was almost four times higher for Black residents:** Zachary Hansen, "American Dream for Rent: Investors Zero In on Black Neighborhood," *Atlantic Journal-Constitution*, February 9, 2023, https://www.ajc.com/american -dream/race-homeownership.

130 **bullish on the future of SFR:** Brien and Wiel, *The Big Long*, 179.

130 **"These groups have a mandate":** "Investing, Renting and the Future of the American Dream," Mike Simonson, Altos Research, *Top of Mind* podcast, undated, https://blog.altosresearch.com/mynds-doug-brien-on-investing -renting-and-the-future-of-the-american-dream.

130 **40 percent of single-family rentals:** Carlos Waters, "Wall Street Has Purchased Hundreds of Thousands of Single-Family Homes since the Great

Recession. Here's What That Means for Rental Prices," CNBC, February 22, 2023, https://www.cnbc.com/2023/02/21/how-wall-street-bought-single-family-homes-and-put-them-up-for-rent.html.

131 **"We want to get to one million homes"**: Shawn Tully, "Meet the A.I. Landlord That's Building a Single-Family-Home Empire," *Fortune*, June 21, 2019, https://fortune.com/longform/single-family-home-ai-algorithms.

131 **a deficit of at least four million homes**: Sami Sparber, "America's Housing Shortage Explained in One Chart," *Axios*, updated December 16, 2023, https://www.axios.com/2023/12/16/housing-market-why-homes-expensive-chart-inventory.

131 **built-for-rent activity skyrocketed**: Alexander Hermann, "8 Facts about Investor Activity in the Single-Family Rental Market," Housing Perspectives, Joint Center for Housing Studies, July 18, 2023, https://www.jchs.harvard.edu/blog/8-facts-about-investor-activity-single-family-rental-market.

131 **the national homelessness rate increased by 12 percent**: *The 2023 Annual Homelessness Assessment Report (AHAR) to Congress*, U.S. Department of Housing and Urban Development, December 2023, https://www.huduser.gov/portal/sites/default/files/pdf/2023-AHAR-Part-1.pdf.

132 **paying more than half of their income**: Jason DeParle, "Record Rent Burdens Batter Low-Income Life," *New York Times*, December 11, 2023, https://www.nytimes.com/2023/12/11/us/politics/rent-burdens-low-income-life.html.

132 **28 percent of Black investors**: "Ariel-Schwab Black Investor Survey (2022)," Schwab Moneywise, Charles Schwab, https://www.schwabmoneywise.com/tools-resources/ariel-schwab-survey-2022.

132 **buying a house would become harder**: "2023 Gen Z and Millennial Survey," Deloitte, https://www2.deloitte.com/content/dam/Deloitte/si/Documents/deloitte-2023-genz-millennial-survey.pdf.

133 **$1.7 billion worth of crypto was stolen**: Chainalysis Team, "Funds Stolen from Crypto Platforms Fall More Than 50% in 2023, but Hacking Remains a Significant Threat as Number of Incidents Rises," Chainalysis, January 24, 2024, https://www.chainalysis.com/blog/crypto-hacking-stolen-funds-2024.

134 **"barrier to entry for real estate investing"**: Jerry Chu, Lofty, undated, retrieved, August 5, 2023, https://www.lofty.ai/team.

134 **"making property ownership inclusive"**: "We're Making Real Estate Investing Available . . . ," Landa, https://www.landa.app/about.

134 **"radically accessible real estate investing":** "A Scientific Approach to Single-Family Rentals," Roofstock, undated, retrieved August 5, 2023, https://www.roofstock.com/.

134 **15,000 homes on its platform:** Amy Rose Dobson, "Proptech Greets Another Unicorn with Roofstock's $240 Million Series E Fundraise," *Forbes*, March 11, 2022, https://www.forbes.com/sites/amydobson/2022/03/11/proptech-greets-another-unicorn-with-roofstocks-additional-240m-series-e/?sh=2753e20172e1.

135 **the whiff of "predatory inclusion":** Keeanga-Yamahtta Taylor, *Race for Profit* (Chapel Hill: University of North Carolina Press, 2019), https://uncpress.org/book/9781469663883/race-for-profit.

135 **"on predatory and exploitative terms":** Taylor, *Race for Profit*, 18.

136 **legal rights afforded through property law are upended:** R. Wilson Freyermuth, Christopher K. Odinet, and Andrea Tosato, "Crypto in Real Estate Finance," University of Missouri School of Law, Legal Studies Research Paper No. 2022-13, November 7, 2022, updated January 30, 2024, https://papers.ssrn.com/sol3/papers.cfm?abstract_id=4268587.

136 **extremely murky legal gray area:** "Use of Tokens or Other Similar Products in Real Property Transactions Committee," Uniform Law Commission, undated, https://www.uniformlaws.org/viewdocument/jeburpa-report-on-the-use-of-tokens?CommunityKey=cccbc3bb-005e-4bc3-869f-38ff3da99186&tab=librarydocuments.

136 **"level of education in the public is pretty low":** Rebecca Szkutak, "Startups Are Looking to Fractionalize Real Estate Assets, but Should They?" *Forbes*, March 1, 2022, https://www.forbes.com/sites/rebeccaszkutak/2022/03/01/startups-are-looking-to-fractionalize-real-estate-assets-but-should-they.

136 **fourteen properties in which they had a stake:** "My Experience with Lofty after 2 Years," r/lLoftyAI, https://www.reddit.com/r/LoftyAI/comments/17mddw9/my_experience_with_lofty_after_2_years (inactive).

137 **DAO is tasked with making some decisions about the property:** Max Ball, "Token Holders Make All Property Decisions via Our Governance System," Lofty, undated, https://learn.lofty.ai/en/articles/6140351-token-holders-make-all-property-decisions-via-our-governance-system.

137 **Token holders are also allowed:** David Ingram, "Crypto Boom Opens Door to a New Class of Landlords," April 24, 2022, NBCNews.com, https://www.nbcnews.com/tech/crypto/crypto-real-estate-investment-landlords-rcna20029.

140 "combined all the responsibilities of homeownership": Ta-Nehisi Coates, "The Case for Reparations," *The Atlantic*, June 2014, https://www.theatlantic .com/magazine/archive/2014/06/the-case-for-reparations/361631.

140 Black families were stripped of wealth: "Rent to Own the American Dream," TechEquity Collaborative, November 2022, https://techequitycol laborative.org/wp-content/uploads/2022/11/Rent-to-Own-the-American -Dream.pdf.

140 ten million Americans have used an RTO product: "Millions of Americans Have Used Risky Financing Arrangements to Buy Homes," Pew Trusts, issue brief, April 14, 2022, https://www.pewtrusts.org/en/research-and-analysis /issue-briefs/2022/04/millions-of-americans-have-used-risky-financing -arrangements-to-buy-homes.

142 purchase the home from Divvy for almost $380,000: Matthew Gold- stein, "Divvy Wants to Make Rent-to-Own Deals Easy. Many Customers Find Them Hard," *New York Times*, August 1, 2023, https://www.nytimes .com/2023/08/01/business/divvy-homes-housing-rent.html.

142 "rainwater often seeped in": Goldstein, "Divvy Wants to Make Rent-to -Own Deals Easy."

142 "Our responsibility in this situation": Ainslie Harris, "Inside the Rent-to -Own Startup That's Putting Aspiring Homeowners in Financial Jeopardy," *Fast Company*, October 24, 2022, https://www.fastcompany.com/90795531 /divvy-real-estate-venture-capit.

143 a consistent growth path: "Historical US Home Prices: Monthly Median from 1953–2024," DQYDJ, https://dqydj.com/historical-home-prices.

143 the company had filed eviction notices on 190: Goldstein, "Divvy Wants to Make Rent-to-Own Deals Easy."

144 just under 50 percent: Divvy has claimed that this rate is higher than their competitors, but it is difficult to verify that claim.

145 "When your customers think": "A Novel Path to Home Ownership," Adena Hefets, Divvy Homes, a16z, undated, https://www.youtube.com/watch?v =tMvvrMtV9TU&t=705s.

145 "Nobody takes good care of a rental": "When Software Eats the Real (Estate) World," Ramp.com, undated, a16z, https://www.youtube.com/watch?v=IR PH3K1GXj0&t=1307s.

146 "Just stay in that house": Harris, "Inside the Rent-to-Own Startup That's Putting Aspiring Homeowners in Financial Jeopardy."

158 combining some of the riskiest elements: *Fortune* editors, "Founder Adam Neumann Explains Why Marc Andreessen Invested $350 Million in Flow,

His New Company That Sounds a Lot like WeWork," *Fortune*, July 13, 2023, https://fortune.com/2023/07/13/adam-neumann-why-marc-andreessen -invested-flow-wework.

158 **"consumer-facing residential brand"**: Clint Rainey, "Adam Neumann Talked about Flow for a Full Hour, and We Still Don't Know What It Is," February 8, 2023, https://www.fastcompany.com/90847220/adam-neumann-a16z -flow-startup-real-estate-explained.

158 **"Adam is a visionary leader"**: Marc Andreessen, "Investing in Flow," Andreessen Horowitz, August 15, 2022, https://a16z.com/2022/08/15/investing -in-flow.

5

THE PATHBREAKERS

161 **VC consistently outperforms the S&P 500**: Maureen Austin, David Thurston, and William Prout, "Building Winning Portfolios through Private Investments," Cambridge Associates, August 2021, https://www.cambridge associates.com/insight/building-winning-portfolios-through-private -investments.

161 **top twenty Silicon Valley firms**: Andy Rachleff, "Demystifying Venture Capital Economics, Part 1," Wealthfront, June 19, 2014, https://www.wealth front.com/blog/venture-capital-economics.

162 **90 percent don't perform any better than the stock market**: Michael D. McKenzie and William Janeway, "Venture Capital Funds and the Public Equity Market," *Accounting and Finance* 51, no. 3 (September 2011), https:// www.researchgate.net/publication/227370556_Venture_capital_funds _and_the_public_equity_market.

162 **fifteen companies will be responsible for 97 percent**: Nicole Perlroth, "Venture Capital Firms, Once Discreet, Learn the Promotional Game," *New York Times*, July 22, 2012, https://www.nytimes.com/2012/07/23/business/ven ture-capital-firms-once-discreet-learn-the-promotional-game.html.

162 **5 percent of venture-backed companies**: Abe Othman and Matthew Speiser, "What Percentage of AngelList Seed-Stage Startups Become Unicorns?" AngelList, July 16, 2021, https://www.angellist.com/blog/angellist -unicorn-rate.

165 **"All of our companies"**: Astrid Scholz, "Where Unicorns Fear to Tread— Building Businesses That Are Better for the World," Medium.com, February

2, 2020, https://medium.com/zebras-unite/where-unicorns-fear-to-tread
-building-businesses-that-are-better-for-the-world–35190e632c9e.

166 **"we create a new model":** Zebras Unite, "Sex & Startups," Medium.com,
February 16, 2016, https://medium.com/@sexandstartups/sex-startups-53f2f
63ded49#.6xc7ss1ew.

170 **three thousand venture-backed startups went out of business:** Erin
Griffith, "From Unicorns to Zombies: Tech Start-Ups Run Out of Time and
Money," *New York Times*, December 7, 2023, updated January 9, 2024, https://
www.nytimes.com/2023/12/07/technology/tech-startups-collapse.html.

170 **the highest failure rate:** Mark Sullivan, "Venture-Backed Startups Are
Failing at Record Rates," *Fast Company*, August 4, 2023, https://www.fast
company.com/90933648/venture-backed-startups-are-failing-at-record
-rates.

170 **"We wanted founders to see":** Bryce Roberts, "V3," Medium.com, January
1, 2019, https://medium.com/strong-words/v3–e9542ba9aeeb.

171 **"We have now reached a point":** Evan Armstrong, "Venture Capital Is Ripe
for Disruption," Every.to, October 27, 2022, https://every.to/napkin-math
/venture-capital-is-ripe-for-disruption.

172 **invest in lots of relatively low-budget movies:** Bryce Roberts, "The Low-
Budget Startup," Medium.com, August 17, 2023, https://bryce.medium.com
/the-low-budget-startup-f7d9ed7f139a.

173 **Aderinkomi, a serial entrepreneur:** "Second Time's the Charm," Carlson
School of Management, University of Minnesota, March 4, 2020, https://
carlsonschool.umn.edu/news/second-times-charm.

174 **"Damn, I wish I would have found":** Thompson Aderinkomi, "Indie and Me,"
Medium.com, January 1, 2019, https://medium.com/@ThompsonAder/indie
-and-me–86b2e0abeb5c.

177 **"I wish every founder":** Aderinkomi, "Indie and Me."

177 **"The shift in strategy":** Bryce Roberts, "The End of Indie," Medium.com,
March 2, 2021, https://bryce.medium.com/the-end-of-indie–6e1b92d90b09.

178 **"We have an opportunity to change":** Jane Thier, "ButcherBox's Founder
Was Fired from His First CEO Gig after 'Losing Everyone's Money.' Here's
How He Learned from His Mistakes to Build a $500 Million Meat Subscrip-
tion Empire," *Fortune*, February 4, 2024, https://fortune.com/2024/02/04
/mike-salguero-butcher-box-first-million.

179 **when he started ButcherBox:** "ButcherBox: Mike Salguero," *How I Built This*
(podcast) with Guy Raz, undated, https://podcasts.apple.com/us/podcast
/butcherbox-mike-salguero/id1150510297?i=1000580453005.

181 **"I don't have investors":** "Why Entrepreneurs Don't Need Venture Capital

to Scale," *Harvard Business Review, IdeaCast,* podcast episode 920, June 27, 2023, https://hbr.org/podcast/2023/06/why-entrepreneurs-dont-need -venture-capital-to-scale.

184 **"You guys should take the money":** Josie Sivigny, "We Turned Down a 'Once in the Lifetime Opportunity' to Build the Company of a Lifetime," Tuftand Needle.com, June 9, 2017, https://www.tuftandneedle.com/blogs/culture /we-turned-down-once-lifetime-opportunity.

187 **corporate certification program:** "FAQs: How Did the B Corp Move- ment Start?" https://www.bcorporation.net/en-us/faqs/how-did-b-corp -movement-start/#:~:text=B%20Lab%20was%20founded%20in,tools %2C%20programs%2C%20and%20more.

189 **increased every year:** "Cumulative Amount of Funding Pledged to Kick- starter Projects from July 2012 to January 2024," Statista, January 29, 2024, https://www.statista.com/statistics/310218/total-kickstarter-funding.

189 **"ensure that artificial general intelligence benefits":** "About," OpenAI.com, https://openai.com/about.

191 **"align our corporate governance":** "The Long-Term Benefit Trust," An- thropic, September 19, 2023, https://www.anthropic.com/index/the-long -term-benefit-trust.

191 **transferring all of its voting shares:** Yvon Chouinard, "Earth Is Now Our Only Shareholder," Patagonia.com, undated, https://www.patagonia.com /ownership.

6

THE CHANGEMAKERS

199 **use the complicated regulations:** Arleen Jacobius, "California Funds Feel Private Equity Shock," P&I Online, May 14, 2018, https://www.pionline .com/article/20180514/PRINT/180519947/california-funds-feel-private -equity-shock.

200 **small proportion of most institutional investment:** "The Yale Endow- ment 2020," SquareSpace.com, undated, https://static1.squarespace.com /static/55db7b87e4b0dca22fba2438/t/607e4da7bc999d01b4752ea2 /1618890160689/2020+Yale+Endowment.pdf.

201 **"to advance the public interest":** "The Netgain Partnership Is Advancing the Public Interest in the Digital Age," NetgainPartnership.org, March 2024, https://www.netgainpartnership.org.

201 **"to ensure that AI advances":** "Philanthropies Launch New Initiative to

Ensure AI Advances the Public Interest," Ford Foundation, undated, https://www.fordfoundation.org/wp-content/uploads/2023/10/Philanthropies-Launch-New-Initiative-to-Ensure-AI-Advances-the-Public-Interest-1.pdf.

203 **"Foundations like MacArthur":** John Palfrey, "Aligning Our Investments with Our Mission, Values, and Programs," MacArthur Foundation, September 22, 2021, https://www.macfound.org/press/perspectives/aligning-our-investments-with-our-mission-values-and-programs.

204 **endowment has performed slightly better:** "Mobilizing More for Mission: Re-designing Wallace Global Fund's Endowment," Wallace Global Fund, undated, https://wgf.org/wp-content/uploads/2022/04/mobilizing-more-for-mission.pdf.

205 **a 2022 survey by PwC:** *ESG-Focused Institutional Investment Seen Soaring 84% to US$33.9 Trillion in 2026, Making Up 21.5% of Assets under Management: PwC Report,* December 22, 2022, https://www.pwc.com/id/en/media-centre/press-release/2022/english/esg-focused-institutional-investment-seen-soaring-84-to-usd-33-9-trillion-in-2026-making-up-21-5-percent-of-assets-under-management-pwc-report.html.

205 **"ESG funds would improve their overall returns":** David Stevenson, "LPs Push GPs to Tie Fund Fees to Impact Goals," PitchBook.com, August 21, 2023, https://pitchbook.com/news/articles/PE-limited-partners-impact-fees.

205 **appropriate for private markets, and venture capital:** Johannes Lenhard and Elena Lutz, "What ESG Financing Means for Venture Capital, White Paper #1," Google Docs, 2021, https://drive.google.com/file/d/1HsC22zi8Gh8W7qgbi9ma4ZtYdEe13e8p/view.

209 **overwhelming show of support for Altman:** Deepa Seetharaman, Berber Jin, and Keach Hagey, "OpenAI Investors Keep Pushing for Sam Altman's Return," *Wall Street Journal,* updated November 21, 2023, https://www.wsj.com/tech/openai-employees-threaten-to-quit-unless-board-resigns-bbd5cc86.

211 **imperative to invest in the private sector:** Tom Nicholas, *VC: An American History* (Cambridge, MA: Harvard University Press, 2019).

212 **credits the SBIC program:** "Oral History with William H. Draper III," Computer History Museum, 2019, https://archive.computerhistory.org/resources/access/text/2019/03/102740503-05-01-acc.pdf.

212 **He took $300,000 from the program:** "Sutter Hill Ventures Portfolio Companies," Crunchbase.com, 2024, https://www.crunchbase.com/hub/sutter-hill-ventures-portfolio-companies.

212 **create more economic growth:** Nicholas, *VC: An American History*, fn95.

214 **"are not sufficiently financed by private market investors":** "Biden-Harris Administration to Establish Reforms to Transform Public-Private Investment Program," U.S. Small Business Administration, Press release 23–46, July 17, 2023, https://www.sba.gov/article/2023/07/17/biden-harris-administration-establish-reforms-transform-public-private-investment-program.

215 **fund managers have always charged:** Victoria Ivashina and Josh Lerner, *Patient Capital: The Challenges and Promises of Long-Term Investing* (Princeton, NJ: Princeton University Press, 2019), https://press.princeton.edu/books/hardcover/9780691186733/patient-capital.

218 **violation of antitrust laws:** Matthew Wansley and Samuel Weinstein, "Venture Predation," *Journal of Corporate Law* 813 (2023), May 4, 2023, revised September 30, 2023, https://papers.ssrn.com/sol3/papers.cfm?abstract_id=4437360.

219 **essay extolling Bankman-Fried's genius:** Adam Fisher, "Sam Bankman-Fried Has a Savior Complex—and Maybe You Should Too," Sequoia, September 22, 2022, https://web.archive.org/web/20221027181005/https://www.sequoiacap.com/article/sam-bankman-fried-spotlight.

220 **"[Bankman-Fried] is committed":** Sequoia Capital, Twitter post, November 9, 2022, 8:52 p.m., https://twitter.com/sequoia/status/1590522718650499073.

220 **"potential to stifle innovation":** Brooke Masters, "US Regulators Impose Tougher Disclosure Rules on Private Funds," *Financial Times*, August 23, 2023, https://www.ft.com/content/7b31f328-27b0-4a96-8477-1aa2b6bd3ae4.

221 **sue the SEC:** Paul Kiernan, "Private Equity, Hedge Funds Sue SEC to Fend Off Oversight," *Wall Street Journal*, updated September 21, 2023, https://www.wsj.com/finance/regulation/private-equity-hedge-funds-sue-sec-to-fend-off-oversight-345ce372.

222 **accounted for more than 75 percent:** Josh Lerner and Ramana Nanda, "Venture Capital's Role in Financing Innovation: What We Know and How Much We Still Need to Learn," Harvard Business School, Working Paper 20-131, 2020.

223 **There is a growing consensus:** *From Laggard to Leader*, Report from the Global Financial Markets Center at Duke Law, February 2021, https://web.law.duke.edu/sites/default/files/centers/gfmc/From-Laggard-to-Leader.pdf.

223 **considering new rules:** Jeff Farrah, "SEC Could Pull More 'Unicorns' into

Public Reporting Regime," NVCA, January 28, 2022, https://nvca.org/sec
-could-pull-more-unicorns-into-public-reporting-regime.

CONCLUSION

226 **the number of small businesses created since the 1980s:** Rana Foroohar,
Makers and Takers (New York: Currency, 2016), 10.

INDEX

INDEX

ABOUT THE AUTHOR

CATHERINE BRACY is a civic technologist and community organizer whose work focuses on the intersection of technology and political and economic inequality. She is the founder and CEO of TechEquity, was previously Code for America's Senior Director of Partnerships and Ecosystem, and founded Code for All. During the 2012 election cycle, she was Director of Obama for America's Technology Field Office in San Francisco, the first of its kind in American political history. She is a prolific public speaker for places like Axios and the Personal Democracy Forum. Her TED Talk, "Why Good Hackers Make Good Citizens," has almost one million views. Her work has been highlighted in the *Los Angeles Times*, *The New York Times*, and on NPR.